Seals

Frances Dipper

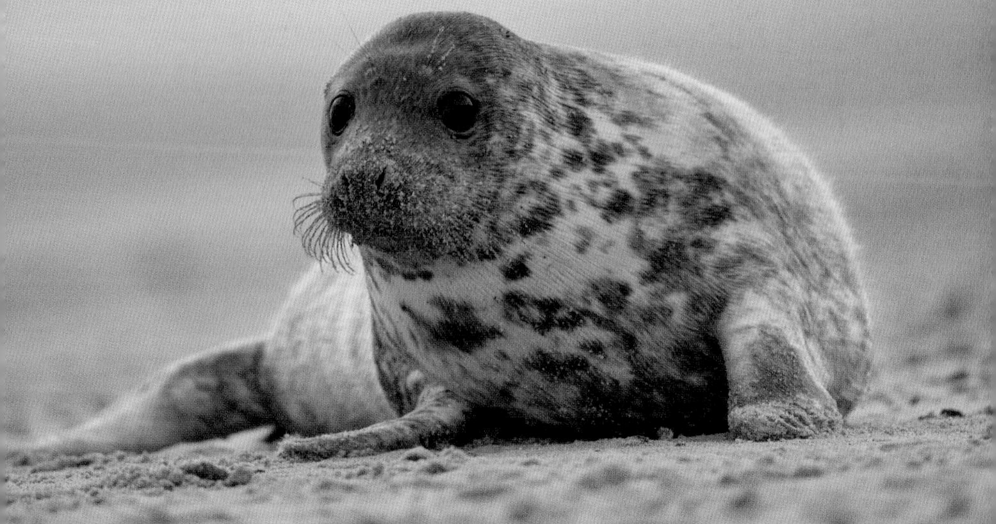

BLOOMSBURY WILDLIFE

LONDON • OXFORD • NEW YORK • NEW DELHI • SYDNEY

BLOOMSBURY WILDLIFE
Bloomsbury Publishing Plc
50 Bedford Square, London, WC1B 3DP, UK

BLOOMSBURY, BLOOMSBURY WILDLIFE and the Diana logo are trademarks of Bloomsbury Publishing Plc

First published in United Kingdom 2021

Copyright © Frances Dipper, 2021
Copyright © 2021 photographs and illustrations as credited on page 127

Frances Dipper has asserted her right under the Copyright, Designs and Patents Act, 1988, to be identified as Author of this work

For legal purposes the image credits on page 127 constitute an extension of this copyright page

All rights reserved. No part of this publication may be reproduced or transmitted in any form or by any means, electronic or mechanical, including photocopying, recording, or any information storage or retrieval system, without prior permission in writing from the publishers

Bloomsbury Publishing Plc does not have any control over, or responsibility for, any third-party websites referred to in this book. All internet addresses given in this book were correct at the time of going to press. The author and publisher regret any inconvenience caused if addresses have changed or sites have ceased to exist, but can accept no responsibility for any such changes

A catalogue record for this book is available from the British Library

Library of Congress Cataloguing-in-Publication data has been applied for

ISBN: PB: 978-1-4729-7162-3; ePub: 978-1-4729-7161-6; ePDF: 978-1-4729-7163-0

2 4 6 8 10 9 7 5 3 1

Design by Rod Teasdale
Printed and bound in India by Replika Press Pvt. Ltd.

To find out more about our authors and books visit www.bloomsbury.com and sign up for our newsletters

Published under licence from RSPB Sales Limited to raise awareness of the RSPB (charity registration in England and Wales no 207076 and Scotland no SC037654).

For all licensed products sold by Bloomsbury Publishing Limited, Bloomsbury Publishing Limited will donate a minimum of 2% from all sales to RSPB Sales Ltd, which gives all its distributable profits through Gift Aid to the RSPB.

Contents

Meet the Seals

Twisting and turning effortlessly in their underwater world, seals are champion swimmers, the sleekest and most agile of all marine mammals. Their aquatic acrobatics allow them to sneak around rocks, appear through curtains of seaweed and swim fast enough to catch a wide variety of fish – their favourite food. Seals spend most of their time underwater, invisible to us as they hunt, explore and play near the coast. It is only when they surface for a breath or during the short times they spend on land that we have the chance to see them.

The seals and their close relatives, the sea lions and Walrus (*Odobenus rosmarus*), together form a group called Pinnipedia, or the pinnipeds. This translates roughly from the Latin as 'having feet as fins', which is indeed what these marine mammals have – two pairs of large fins, more usually called flippers. These and their streamlined body are what make them such excellent swimmers, but they also make them rather slow and clumsy on land. If seals are disturbed when resting on the seashore, they will slip quickly back into the water, where they are safer and feel much more at home. Once there, curiosity will often overcome them and they will bob to the surface, craning their head up and around to peer at boats and strange two-legged humans.

Above: South American Sea Lions (*Otaria flavescens*) hauled out on rocks in Patagonia, South America.

Pinnipeds are one of three main groups of marine mammals, the other two being the cetaceans (whales, dolphins and porpoises) and sirenians (dugongs and manatees). The Sea Otter (*Enhydra lutris*) and the Polar Bear (*Ursus maritimus*) are also classed as marine mammals. These groups of animals are not necessarily closely related, but are defined as true marine mammals because they get all their food from the sea (although Polar Bears also scavenge on land). All mammals, wherever they live, have two things in common: they feed their young milk, and they have hair – even if, as in cetaceans, it is rather sparse. Seals grow a thick fur and fatty blubber that helps them keep warm.

Opposite: Resting on remote and undisturbed shores, Common Seals are a picture of relaxed contentment.

Shaped for swimming

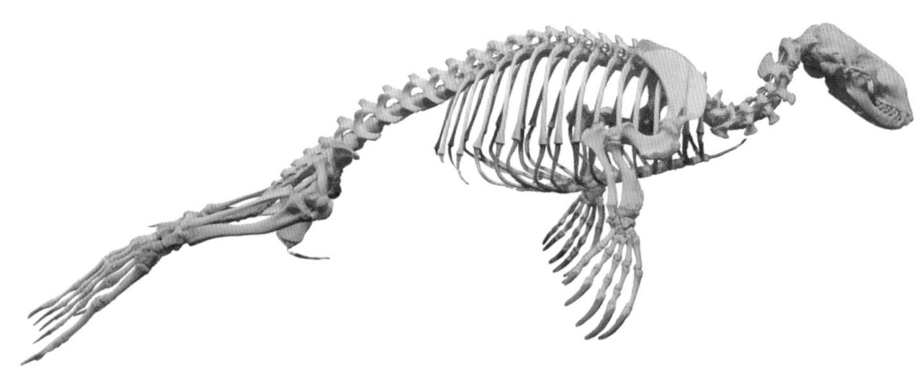

Above: A seal's skeleton is designed for flexibility. Unlike most mammals, they do not have a clavicle (collar bone).

Below: Seals have a similar streamlined shape to fast predatory fish such as Bigeye Trevally (*Caranx sexfasciatus*). These, and top speeders like billfishes, fold their dorsal and pectoral fins away to reduce drag at top speeds.

Swimming through water is hard work, but streamlining can help reduce drag. The most efficient body shape for aquatic animals is a cylinder that tapers at both ends, which is why tuna and other fast predatory fish are this shape. Seals have a similar torpedo-like body with very few projecting parts that might slow them down. Their limbs are shortened and only the flippers are visible from the outside. There are no visible mammary glands in females, as these are internal and the teats are kept turned in until nuzzling by newborn pups pops them out. Similarly, the male seal's sex organs are internal. This is in contrast to terrestrial mammals, in which the male sex organs are carried externally to keep them cool (overheating is not a problem in marine environments). While most land mammals (including humans) have obvious external ears, seals either have very small ear flaps or just a small hole on the side of the head.

Native species

Our coastal waters are home to two species of seal, the Grey Seal (*Halichoerus grypus*) and the Common or Harbour Seal (*Phoca vitulina*).

Grey Seal

Grey Seals live in the North Atlantic Ocean, and about 34 per cent of the world population is found all around the coasts of the British Isles, but especially in the Western Isles (Outer Hebrides) and the Orkney Islands off Scotland. Smaller numbers live around Scandinavia and in the Baltic Sea, around Iceland and along the coastline of north-eastern North America. Unlike Common Seals, Grey Seals are happy on wild, wave-tossed shores. The latest UK non-pup population estimate (2017) is 150,000 and the estimated world population of mature individuals (2016) is 316,000.

In profile, Grey Seals have a long muzzle (the nose and mouth). This is flat on top in females and convex or slightly humped in males, and is often called a 'Roman

Below: Their large size and strength mean that adult Grey Seals are not bothered by waves and surf.

Above: Viewed head-on, the wide-set nostrils of a Grey Seal are often described as two rather parallel slits.

Below: Resting Grey Seals often lie on their side, exposing their belly to the warm sun. Or they lie belly-down raising their head and rear flippers.

nose'. The eyes are set around halfway between the back of the head and the nose, and the two nostril slits are almost parallel (these are best seen when the animal is peering directly at the observer and has its nostrils closed, perhaps just before it disappears underwater).

Young Grey Seal pups (as the babies are called) are easy to distinguish because they are born snow white. As adults, Grey Seals have messy, irregular spots and are noticeably darker on the back than the belly (this is especially true of

Left: Seen in profile, the long snout of a Grey Seal is a distinctive feature. The small hole behind and just above the eye level is the entrance to the ear canal.

males) – although note that the fur coat, or pelage of any seal may vary depending on where they live, whether they have moulted recently, how old they are, and whether it is wet or dry. The adults are also quite large, growing to around 2m (6.5ft) in length. When hauled out on land with their head and tail lifted, they look a bit like a flat-bottomed boat – a clear contrast to the 'U' shape often adopted by Common Seals. This may be because, owing to their larger size and bulk, Grey Seals are less flexible.

Below: The white, fluffy lanugo coat of a baby Grey Seal gives them a vulnerable and appealing appearance, but they have incredibly sharp teeth!

Common Seal

Confusingly, Common Seals are much less common than Grey Seals in our home waters. However, they also live in the Arctic and North Pacific oceans, as well as the North Atlantic, and so are much more widespread. There are

Right: Seen in close-up, the V-shaped nostrils of a Common Seal show up clearly. The snub-nosed look and eyes close to the snout help confirm its identity.

Below: Sunbathing Common Seal-style involves raising head and rear flippers to gain maximum warmth all over.

about 45,100 (2017) around UK shores, with 80 per cent of these in Scotland; worldwide there are at least 315,000 mature individuals. To add further confusion, Common Seals are usually called Harbour Seals everywhere outside the UK. This is because they prefer to live in sheltered places such as bays, sea lochs and estuaries.

A Common Seal in profile has a much stubbier, dished muzzle compared to a Grey Seal, with a relatively distinct forehead. Its eyes are much closer to its nose than in the Grey Seal, and its nostril slits form a distinct 'V' shape. Unlike the snow-white Grey Seal newborns, Common Seal pups are usually a spotty grey, and the adults have a neater appearance with spots that are all much the same size.

Common Seals rarely grow larger than about 1.5m (4.5ft) in length. When on land, they frequently lift up both their head and tail-end to form a rough 'U' shape. This may help them to warm up quickly after a cold swim by exposing more of their skin to the sun's rays. They also react in this way to chilly water splashing up the beach as the tide comes back in.

Below: A pale, spotted coat is typical of many Common Seal pups, but others are born with a very dark coat.

What's the difference?

	Grey Seal	Common Seal
Head shape	'Roman nose' – long muzzle, flat on top in females and convex or slightly humped in males	Stubbier, dished muzzle with a relatively distinct forehead
Eye position	Set around halfway between back of head and nose	Much closer to nose
Nostril shape	Almost parallel	Distinctively V-shaped
Newborn coloration	Snowy white	Spotty grey
Adult coloration	Messier, irregular spots and darker on back than belly (especially males)	Neater appearance than Greys, with similar-sized spots
Adult size	Length 2m (6ft 7in)	Length 1.5m (4ft 11in)
Haul-out body silhouette	Like a flat-bottomed boat	U-shaped

Common or Harbour Seal range

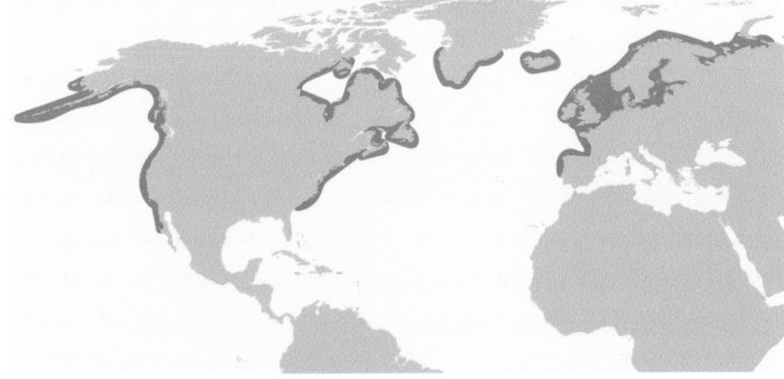

Grey or Gray Seal range

Below: Blakeney Point in Norfolk, UK, hosts England's largest breeding Grey Seal colony, with 3,000 pups in 2018. Grey Seal numbers here and worldwide are increasing.

Occasional visitors

Most seals prefer to live in cold water and are well equipped to survive and thrive in winters with ice and snow. However, seals living in Arctic and subarctic parts of the North Atlantic do sometimes stray further south into Scotland and even England, Wales and Ireland. Four of the six seal species that live in Arctic waters have been recorded as vagrants (meaning well outside their normal range) around British coastlines, as has the closely related Walrus. These rarities occasionally stay in one place for several days or even weeks, attracting many wildlife enthusiasts – just as some birdwatchers will travel large distances to see a rare avian vagrant.

Ringed Seal

This pretty seal is marked with irregular off-white rings. It is the most common seal in the Arctic Ocean and there is also an isolated population in the landlocked Baltic Sea. Some Common Seals have similar markings, so distinguishing a Ringed Seal (*Pusa hispida*) with certainty can be difficult without a close look. Vagrants have been recorded most frequently in Shetland and down the east coast of the UK, but also as far south as the Azores and mainland Portugal.

Bearded Seal

With its elegant moustache of long bristles, this military-looking seal should be easy to spot. Its front flippers are often squared off at the ends like a spade. The Bearded Seal (*Erignathus barbatus*) is the largest of the northern true seals and is found all around the edge of the Arctic Ocean. Vagrants have been spotted on the Northumbrian coast.

Harp Seal

Adult 'Harps' (*Pagophilus groenlandicus*) have a black face and a pale silvery or light grey body with a large black band along each side that joins near the head to form a horseshoe shape. This species is the most abundant seal in the northern hemisphere and has a wide distribution in the northern North Atlantic and adjoining parts of the Arctic Ocean. Harp Seals are capable of long migrations and so it is not surprising that vagrants occasionally turn up in the UK, especially along North Sea coasts. In 2008, an individual was spotted just inside the Mediterranean Sea, the first record that far south. Sadly, however, it did not survive.

Hooded Seal

Picking out adult male Hooded Seals (*Cystophora cristata*) is not difficult because they have an inflatable red balloon-like extension of their nose on top of their head and can blow out another bright red balloon from their nostril – all to see off rival males during the mating season. They live on ice in the Arctic Ocean and in the far north of the North Atlantic Ocean. Vagrants turn up all around the British Isles and even off north-west Africa, although sightings are very rare this far south.

Walrus

Spotting a Walrus in British waters is extremely unusual as they rarely stray south of the Arctic – there have been only around 20 reliable records within recent history. Most of the records are from Scotland. One individual, nicknamed Wally the Walrus, turned up in Orkney in March 2018 and then moved to the west coast of Scotland. Since then, he is thought to have returned to the Arctic.

Wally the Walrus

Perhaps the most famous pinniped to visit Britain was a Walrus called Wally, like his 2018 namesake. In 1981, this youngster turned up near Skegness in Lincolnshire on the east coast of England, then swam into The Wash estuary and followed a barge up the River Ouse to King's Lynn. Near Downham Market, 24km (15 miles) upriver, he caused a sensation when he hauled out onto the bank. The next day Wally escaped all attempts to catch him, swam back downriver and out into The Wash, and returned to almost the same spot near Skegness where he was originally sighted. Following his celebrity status, Wally was finally captured and flown to Greenland to join other resident Walruses. However, he was obviously a seasoned traveller, as a Walrus with a broken left tusk just like his was spotted off the coast of Norway a few months later.

Seal origins and classification

All pinnipeds, including our own Grey and Common Seals, belong to the Carnivora, a group of meat-eating (or in this case fish-eating) mammals, which also includes otters, bears, cats and dogs. Pinnipeds are divided into three families: the true seals (Phocidae), to which both the Grey Seal and Common Seal belong; the sea lions and fur seals (Otariidae); and a tiny family with just one species, the Walrus (Odobenidae).

All three pinniped families are now thought to have descended from a group of land-dwelling carnivorous ancestors, which probably originally lived in fresh water. In 2009, a 20 million-year-old fossilised skeleton of an animal, later named *Puijila darwini*, was found in a dried-up lake in the Arctic. It had a skull like that of a seal and a body, legs and tail like those of an otter. Although this extinct species is not a direct ancestor of the pinnipeds, it has some of the features that such an ancestor must have had. True seals (phocids) appeared between 19 million and 14 million years ago, and eared seals and walruses about 19 million years ago. Their closest living relatives today are bears and mustelids, a group that includes otters.

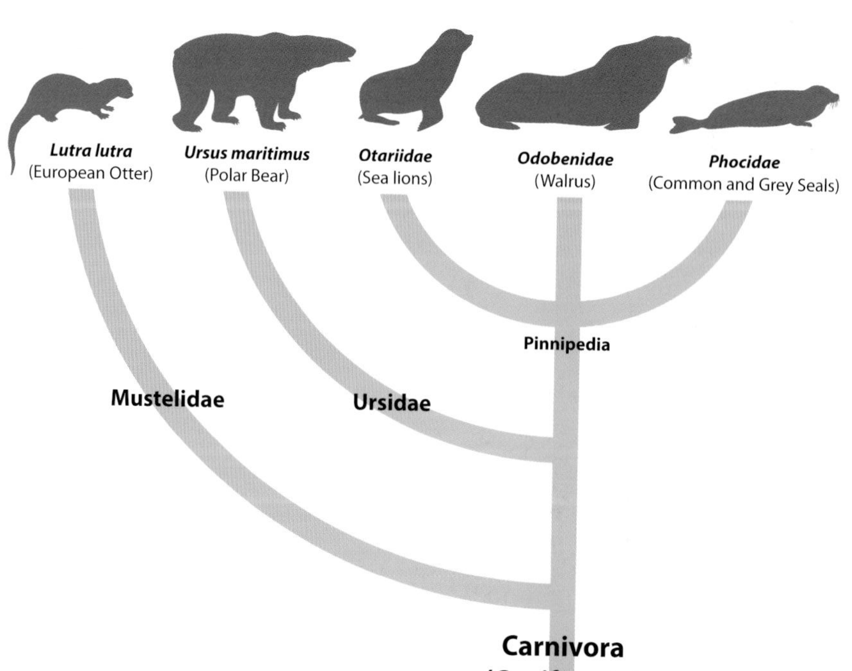

Seals Around the World

Today, there are 33 living species of pinnipeds worldwide, although this figure may increase if genetic studies prove that some subspecies are distinct enough to be classified as true species. Almost all live in the sea, but some have moved into freshwaters. Most are in cold polar regions or cool temperate waters, and only a few make their home in tropical and subtropical seas. That is not surprising since, with their insulating fur and thick blubber, pinnipeds are well suited to living in cool environments.

Less than a century ago, there were 35 living species of pinnipeds, but sadly two seals have since gone extinct. The Caribbean Monk Seal (*Neomonachus tropicalis*) was unusual because it lived in warm tropical waters. The last confirmed sighting was in 1952, but it was already rare a hundred years before that, having been hunted relentlessly for the oil extracted from its blubber. In the Pacific Ocean, the Japanese Sea Lion (*Zalophus japonicus*) is also extinct, having disappeared sometime in the 1950s.

Today, the two rarest pinnipeds are the Mediterranean Monk Seal (*Monachus monachus*) and the Hawaiian Monk Seal (*Neomonachus schauinslandi*), both of which are classified as Endangered on the International Union for Conservation of Nature (IUCN) Red List. The Galápagos Fur Seal (*Arctocephalus galapagoensis*), Australian Sea Lion (*Neophoca cinerea*) and Galápagos Sea Lion (*Zalophus wollebaeki*) are also classed as Endangered but their numbers are not nearly as low as those of the two monk seal species. The most numerous species is the Crabeater Seal (*Lobodon carcinophaga*). It is very difficult and expensive to survey the numbers of this species, which lives on the Antarctic ice pack in the Southern Ocean, but surveys in the late 1990s indicated there were at least 8–10 million individuals and perhaps considerably more.

Above: A Mediterranean Monk Seal relaxing in shallow water. This is one of the rarest pinniped species in the world.

Opposite: Crabeater Seals are one of the most numerous pinniped species in the world. They have a circumpolar distribution around the Antarctic continent.

True seals

All the seals found in the UK and Europe are true or earless seals, and are members of the family Phocidae. There are 18 species of true seals worldwide, including the two largest pinnipeds of all, the elephant seals. The Southern Elephant Seal (*Mirounga leonina*) is found in the cold Southern Ocean, along with four other true seals. The Northern Elephant Seal (*Mirounga angustirostris*) lives in the north-east Pacific Ocean. The other 12 true seals also live in the northern hemisphere, mostly in the Arctic and subarctic.

Although they are often called earless seals, the members of this family do have ears and excellent hearing – they just have no external ears. A tiny opening, visible on each side of the head behind the eyes, leads into the auditory canal. The other main characteristic of true seals is that they cannot turn their hind flippers forwards, and so must trail their hind end and shuffle or galumph along on land.

Below: Common and other earless seals can close off their ear holes when they submerge. Specialised tissues in the middle ear help pressure equalisation when diving.

Cold-water giants

Everything about elephant seals is giant. Adult bull (male) Southern Elephant Seals can weigh nearly 4 tonnes (4.4 tons), almost as much as an Asian Elephant (*Elephas maximus*). Both species of elephant seals have a huge nose that hangs over the mouth like a miniature trunk, and in the northern species it dangles down as much as 25cm (10in) below the lower jaw. This proboscis, as it is called, can be inflated and acts like a megaphone, allowing rival males to roar loudly at one another. The louder the roar, the stronger the male, or so it seems – dominant males can often chase off a rival just by shouting at it.

Above: Like all true seals, the Ross Seal has no external ears and its rear flippers trail behind it. This seal lives on the ice in the Southern Ocean.

Below: Only male elephant seals have the enlarged, trunk-like nose that gives them their name. It develops gradually, starting off as a fleshy lump behind the nostrils in subadults.

Right: Rival Southern Elephant Seal bulls inflict nasty bites, rearing up and swinging their massive heads at each other. The loser is usually the smaller seal, which is ignominiously bitten on the rear end as it finally makes its escape.

Below: Female Elephant Seals are much, much smaller than the giant males.

Females are dwarfed by the males and are certainly bossed around by them. A large dominant male keeps a harem of females in his beach territory and chases off all other rivals. The bulls are not very good parents either, and sometimes crawl over and crush pups or even actively attack them. Things are much more peaceful outside the breeding season, when the seals swim large distances and can stay at sea for weeks or months at a time, feeding and fattening themselves up.

Elephant seals have enormous amounts of blubber to keep them warm while they search for their favourite squid and fish dinners. This blubber was once a valuable source of oil, and in the 19th and early 20th centuries both species were hunted intensively – the northern species almost to extinction. Once the hunting stopped, populations recovered and today the species are not under threat from humans. However, no one knows what effect ocean warming may have on them or their prey species.

Tropical island residents

Most seals live in cold water, but Hawaiian Monk Seals are the only true seal species to spend their entire lives in warm tropical waters. They live mostly around the remote Northwestern Hawaiian Islands, although some are also found on the popular holiday destination islands at the south-west end of the island chain. Hawaii is isolated in the middle of the Pacific Ocean and Hawaiian Monk Seals are endemic to the archipelago – that is, they are found nowhere else in the world, as is the case with many other Hawaiian animals and plants. This means that if anything happens to them, there are no other populations to take their place.

Even though the seals live in such a remote location, they were hunted to near extinction in the 19th century. Most of the area in which they live is now protected, both as a UNESCO World Heritage Site and a United States National Monument, which has the tongue-twisting name of Papahānaumokuākea. Despite being protected under US law, only about 1,200 Hawaiian Monk Seals are now left. Great efforts are being made to help these seals and there is now a dedicated Hawaiian Monk Seal hospital, called Ke Kai Ola (which means 'The Healing Sea') located in Kailua on the island of Hawaii, to help sick and injured individuals.

Below: Hawaiian Monk Seals prefer to live on their own. Most are frightened by people and don't like being disturbed.

Freshwater seals

Hidden away in remote southern Siberia in Russia is the oldest and deepest lake in the world. Lake Baikal was formed from an ancient rift valley around 25–30 million years ago and is 1,700m (5,600ft) deep. This remote landlocked lake seems an unlikely place for a seal to live, but it is home to the only truly freshwater seal in the world, the Baikal Seal (*Pusa sibirica*). Around 40,000 years ago, adventurous Ringed Seals are thought to have swum up the river systems that drained from the lake into the Arctic Ocean. Isolated from other populations, these ancestral seals gradually evolved into a new species.

Today, around 80,000–100,000 Baikal Seals live in the lake. There are plenty of nutritious, oil-rich fish for them to feed on, and numerous islands and rocky shores make good haul-out sites. In winter, the lake freezes over and the seals live on the ice but keep access holes ice-free. The pups are also born on the ice, hidden away in lairs dug into overlying snow. In summer, some seals make occasional short forays up rivers that flow into the lake. But as with all seals, there are always a few adventurers,

Below: In summer, Baikal Seals turn rocky islets dark grey as they haul out and lie around in sociable groups.

Above: In winter, Baikal Seals are mostly found alone on the ice, the females with their pups.

and individuals have been recorded several hundred kilometres up the Angara River, today the only large outlet from the lake.

While Baikal Seals are still hunted for their skins, meat and blubber, the numbers taken each year are controlled and relatively small. A much greater threat may be global warming. The pups are born on ice, and if the amount of winter ice decreases, there will be less space for the females to establish breeding territories. Increasing pollution is another major concern.

Other seals do live in freshwater lakes, but these are colonies from species that live mostly in the ocean. Ringed Seals are found throughout the Arctic Ocean, but isolated populations live in Lake Ladoga in Russia and Lake Saimaa in Finland. Caspian Seals (*Pusa caspica*) live only in the Caspian Sea, a landlocked body of brackish (slightly salted) water that was connected to the sea around 5.5 million years ago.

Sea lions and fur seals

If you have ever been to a zoo, then in addition to 'real' Lions (*Panthera leo*), you may well have seen a sea lion – probably a California Sea Lion (*Zalophus californianus*). If you look closely the next time you go, you will see that these clever animals have tiny but clearly visible ears called pinnae, as do fur seals. Sea lions and fur seals all belong to the family Otariidae, commonly known as eared seals, and include 14 species. There are not many obvious differences between sea lions and fur seals. As their name suggests, fur seals have an especially thick, two-layered fur coat to keep them warm, which is just as well since many of them live in very cold places, including the Antarctic. The fur of sleek sea lions is

Below: This inquisitive Australian Sea Lion (*Neophoca cinerea*) can see exactly what is going on around it because it can curl its hind flippers around and under its body, and then push itself up on its long front flippers.

Above: Sea lions and fur seals are called eared seals (Otariidae) because they have an external ear flap at the ear entrance. The flap points downwards to protect the ear underwater.

thinner but still plenty thick enough to keep them warm. Large adult male sea lions usually have a thick mane of shaggy fur on their head and neck, just like their terrestrial namesakes. But then so do some male fur seals – although theirs is not usually quite as magnificent.

On land, eared seals can swing their hind flippers around so that they are facing forwards. This means they can 'walk' or 'run' along using both pairs of flippers – something a true seal cannot do. In the water, eared seals swim using their large front flippers.

In the wild, California Sea Lions are bold hunters and will chase and find a very wide variety of prey. They can outswim fast fish such as mackerel and herring, search rocky crevices for octopuses and dig out flatfish from the seabed. They can also stay underwater for at least 10 minutes at a time, dive to depths of around 300m (980ft) and return straight to the surface without risk of the bends. Thanks to these skills, sea lions have been trained by humans for entertainment and for more serious roles.

Naval friends

Wild California Sea Lions live along the west coast of the USA and Mexico, but are familiar to people all around the world as captive performers in zoos, oceanariums and even circuses. Trained individuals can balance, spin and toss a ball on their nose and clap their front flippers. Their intelligence is also put to more serious use – the US Navy trains California Sea Lions to find and mark the position of lost or test equipment. The sea lion carries a device in its mouth to clamp onto the object, which is then hauled to the surface with an attached rope. This is especially useful in murky or deep water where human divers find it difficult to operate. The sea lions are also trained to find intruders swimming into restricted areas and to find unarmed test ordnance such as mines.

Above: Taken in 1974, this photograph shows a trained sea lion attaching a grabber device to a mock ASROC missile.

Walrus

The fossil record tells us that there were once many walrus species, but today there is only one. This is what is known as a 'relic' species – the only one left from what was once a common group of pinnipeds in prehistoric times.

Walruses are huge animals, are almost bald and have thick, wrinkled skin rather like an elephant's. With their blubbery looks they are certainly not sleek and elegant like seals, but they need these great rolls of fat to keep them warm in icy Arctic weather, as they lack fur. Their blubber layer, which is up to 15cm (6in) thick, also allows them to survive when food is scarce, giving them plenty of supplies for a move to richer feeding areas.

Above: Walruses live in cold Arctic and subarctic regions including Greenland, Alaska, northern Canada and Russia.

Along with true seals, Walruses have no visible ears, and like eared seals they can walk quite well on land – they, too, can tuck their hind flippers forwards under their body. They are the only pinniped to have tusks – huge, extended canine teeth that can grow to 1m (3ft) long. It was these tusks that nearly led to the extinction of the species, as Walruses were hunted relentlessly, especially in the 19th century, for their ivory, skin and oil. If a Walrus breaks its tusks it can survive because it does not need them for feeding (as was once thought), but males with no tusks or broken tusks have a low social status, making it difficult for them to compete for females.

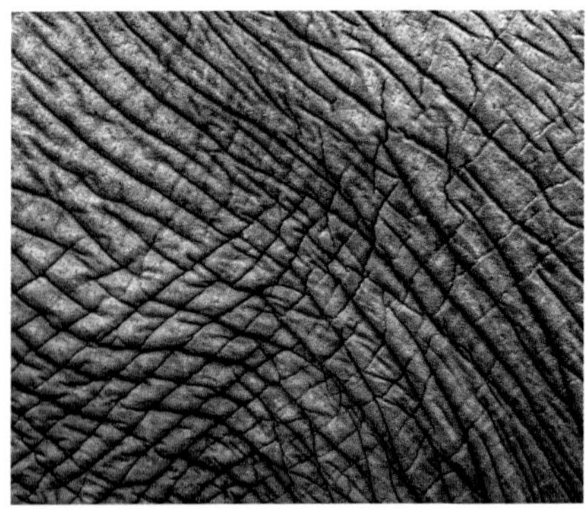

Above: Like elephants (right), Walruses (left) have extremely wrinkled skin.

Below: This specimen in the Horniman Museum in London has no wrinkles because when it was preserved more than a century ago, few people had actually seen a Walrus, and so it was overstuffed by the taxidermist.

Strong tusks can also keep predatory Orcas or Killer Whales (*Orcinus orca*) and Polar Bears at bay, and are used by the heavy pinnipeds to help haul themselves out of the water onto ice floes and to keep breathing holes in pack ice clear.

Walruses forage along the seabed, sucking water into their mouth and then squirting it back out like a high-pressure water pistol to blow away the mud covering bivalve clams, their favourite prey. They then slurp these molluscs back into their mouth. Walrus whiskers, called vibrissae, are sensitive feelers that help the pinnipeds find and select their hidden food. Walruses are extremely noisy and smelly on land, since they haul out in huge, sociable herds and are not fussy about where they leave their urine and excrement.

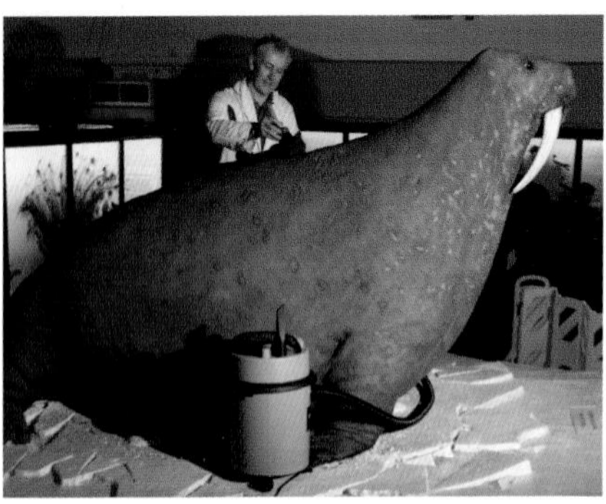

Pink is cool

Even in the Arctic, it can get quite warm if you lie on a beach in the summer sun in a thick coat. Like our own skin, a Walrus's can glow pink after a few hours, but this is not sunburn. In warm weather more blood flows to the surface of the skin, allowing body heat to escape and helping to keep the Walrus cooler. This rosy-pink colour was the inspiration for the Walrus's scientific species name, *rosmarus*.

Seal records

Finding out about seals can be difficult, especially those species that live in remote and cold places. Gathering information about them is therefore an ongoing process and, like athletes, records are often contested or broken as new data are collected. Records of size and other statistics are often exaggerated in the popular press, but below are a few current records at the time this book was published.

Largest Southern Elephant Seal (male) – 5.8m (19ft) long and 4,000kg (8,800lb).

Smallest Baikal Seal – 1.3m (4.3ft) long and 90kg (200lb).

Deepest and longest diver Southern Elephant Seal – more than 2,000m (6,500ft) and 120 minutes.

Most widespread Common Seal – found throughout the northern hemisphere in temperate and polar waters and native to at least 17 countries.

Closest to South Pole Weddell Seal (*Leptonychotes weddellii*) – lives on fast ice all around Antarctica, keeping breathing holes free by regularly bashing the ice as it re-forms.

Most endangered Mediterranean Monk Seal – current total population of 600–700.

Most numerous Crabeater Seal – current population of more than 8 million (although no firm data).

Hungriest Walrus – can eat up to 6,000 shellfish a day.

Fastest weaning period Hooded Seal – mothers feed their pups for an average of four days, the shortest of any mammal.

Most sociable Walrus – herds can number up to 14,000 individuals. Walruses will defend injured individuals and mothers will defend their pups to the death.

Most dangerous Leopard Seal (*Hydrurga leptonyx*) – a huge head and mouth allow it to feed on penguins and young seals. In 2003, an Antarctic researcher was killed by a Leopard Seal, the only documented human fatality.

Above: Leopard Seals patrol the shoreline of penguin colonies waiting for their prey to venture in.

Life on Land

While seals are superbly adapted to living in water, all species need to come ashore to give birth, rest and moult. So seashores, poised between land and sea, are vitally important to them. However, they face many threats here – from the tide of everlasting plastic and fishing debris that washes up, to disturbance and development that make some shores unsuitable or uncomfortable. Nature, too, can make life difficult for seals – storms can erode their preferred beaches and sometimes even carry away vulnerable pups.

Lumbering locomotion

A seal underwater is as graceful as any ballet dancer, gliding along with powerful sideways sweeps of its trailing hind flippers. Out of water, however, those trailing limbs are of little use and the seal must haul itself along, throwing its front flippers forwards, gripping on and then pulling up its rear end. Despite this, a fit and active seal can still outpace most people, humping along like an animated looper caterpillar. Common Seals often haul themselves out of the water onto mud and sand, leaving behind a characteristic wide body scuff with their flipper prints showing up on either side.

Sea lions, fur seals and Walruses can swing their hind limbs forwards under their body, so they lollop along quite quickly using all four flippers. They can also 'sit up' comfortably, which is one of the reasons captive sea lions can be taught tricks, such as catching a ball on their nose.

Opposite: A Grey Seal dragging itself along the shore in a belly crawl at Blakeney, Norfolk, has left distinctive tracks in the sand.

Below: The ability to swing their hind limbs forward means that sea lions can run fast on land, sometimes faster than people.

Quick exit

As a scuba diver, I know just how difficult it can be to get out of the water onto a rocky shore, encumbered by heavy weights and a diving cylinder, especially in a swell and breaking waves. On several occasions I have been forced to clamber ashore over steep, barnacle-covered rocks and been dumped unceremoniously by every wave. Seals have no such problems.

Tipping the scales at 100–300kg (220–660lb), an adult Grey Seal is no lightweight, but its blubber gives it buoyancy. With great skill, a seal chooses a wave on which to surf ashore, then grips the rock with the long, tough claws of its fore flippers. Pressing its heavy body down, it holds on until the next wave lifts it, when it can push forwards up the shore.

Communal living

Out at sea, seals lead very independent lives and usually go off searching for food on their own. In places with a good supply of fish, there might well be quite a few seals hunting, but they don't seem to cooperate or purposefully stay together in a group. On land, it is a different matter. While you may come across a lone seal or a small group on the shore, there are many places where they gather in large numbers. These favoured spots are called haul-outs, and it is not unusual to find tens or hundreds of individuals together. Good haul-out sites have easy access from the sea and comfortable places to rest on the shore. Although Grey and Common Seals have different habitat preferences, they do sometimes share haul-out sites where their ranges overlap.

In theory, seals do not need to come ashore to rest or sleep as they can do both in the water, but lying around on land saves energy and makes for an easier life. Warming up in the sun and resting with other seals, which help keep a lookout for danger, certainly seems

Below: Although unusual, Common and Grey Seals sometimes share haul-out sites, such as this one at Blakeney Point, Norfolk.

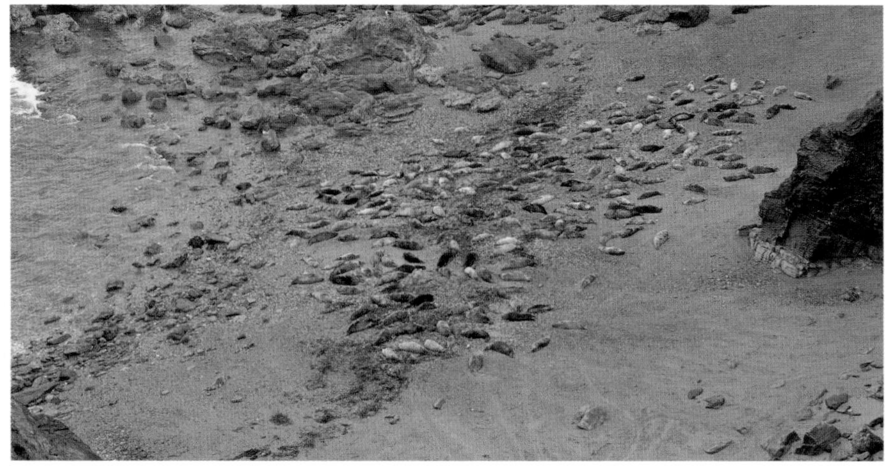

Above: A popular and safe seal haul-out site at Godrevy, Cornwall. Grey Seals are the ones usually seen all round the wave-tossed Cornish coastline.

sensible, and here they are safe from sharks and other potential predators. Just as we revisit favourite cafés or parks, so individual seals seem to favour particular islands, coves or sandbanks, at least for a few days or weeks. We know this because seals have individual markings and so can be recognised through photo identification or from tagging.

Seals vary in the amount of time they spend hauled out, both as individuals and as species. In southern Norway, some patient biologists watched Common Seals throughout 2014 and found that at their sites the pinnipeds spent up to five hours on shore in September, but only an hour or so in February, when temperatures are warmer in the water than out. The ebb and flow of tides, and whether low tides occur at night or during the day, also affect the amount of time seals spend hauled out.

When Grey Seals come ashore to breed, they gather in much larger numbers at special haul-out sites called rookeries. Rookeries are generally used only for breeding and not as normal resting sites. Common Seals also have preferred pupping areas, but are far less fussy. Many seals return to the same breeding beaches in subsequent years. The largest Grey Seal rookeries in the UK are in the Orkney Islands, Outer and Inner Hebrides, and Isle of May in Scotland, the Farne Islands in Northumberland, and Donna Nook in Lincolnshire.

Grey Seal rookeries

Most of the major breeding colonies of Grey Seals in the UK are in Scotland, but there are many other sites with smaller numbers. The blue-shading shows the areas within which about 75% of UK Grey Seal pups are born each year and stay for 2–3 weeks until weaned. Red dots indicate areas with smaller breeding colonies. Around 80% of UK Common Seals live in Scotland; pups swim with their mothers from birth, and nursery areas are more dispersed.

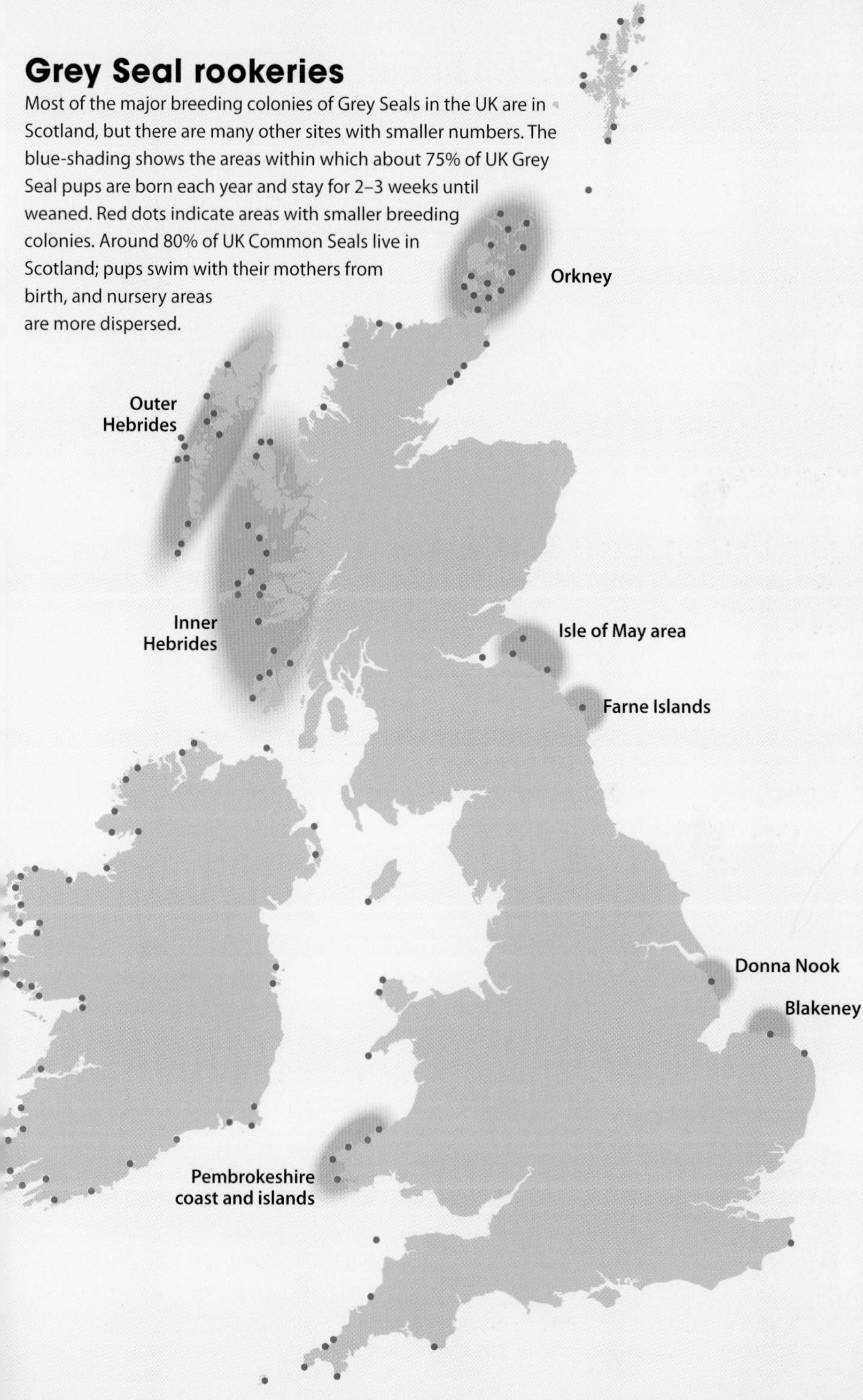

Orkney
Outer Hebrides
Inner Hebrides
Isle of May area
Farne Islands
Donna Nook
Blakeney
Pembrokeshire coast and islands

Social interactions

In spite of hauling out in large groups, Grey and Common Seals are not actually very sociable. Visit any haul-out site and you will see that the seals are spread out, each individual, whether male or female, maintaining its own personal space. That said, Grey Seals are more tolerant of close neighbours than are Common Seals. Getting the best spot first does not mean a seal will keep it, and younger, smaller or more docile individuals may be pushed out by more dominant ones. In rookeries, however, mothers with pups will snarl and threaten much larger males, who soon get the hint that their advances are not welcome.

Aside from mating, only mothers and their pups have any real physical contact or show affection for one another. Grey Seal mothers nuzzle their pups on land and groom them with their front flippers, and Common Seal mothers hold and control their pups in the water to keep them safe. However, mothers of both species more or less ignore their pups once they are weaned. The young seals continue to swim and play together, both in and out of the water, until they become confident enough to go their own way.

Opposite: In contrast to Grey and Common Seals, Walruses are happy to pack close together at their haul-out sites, often without an inch of space between them.

Below: Grey Seals prefer to keep some personal space around them when resting at haul-out sites. Tracks show where individuals have moved away from one another.

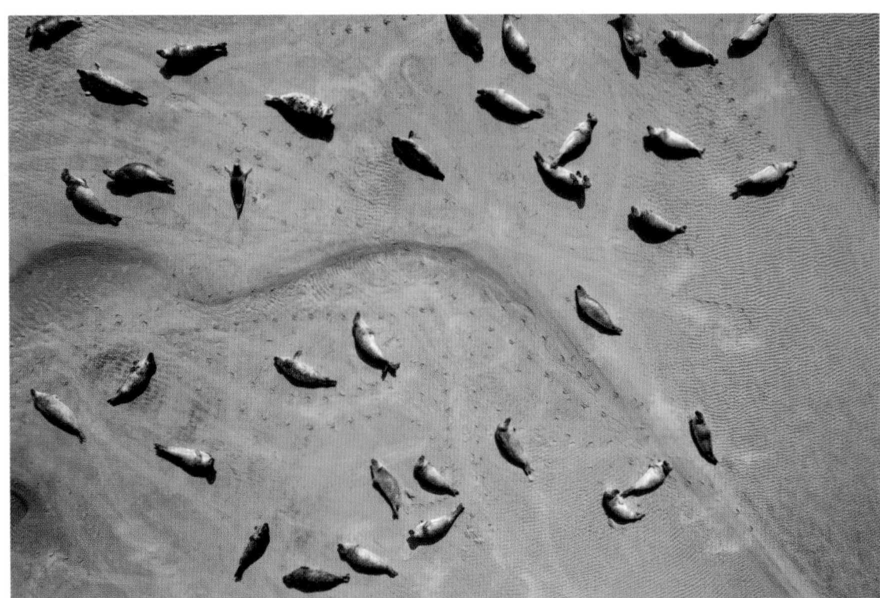

Surprise visits

Seals are adept at finding isolated rocks and sandbanks where they can rest and chill out. However, they have also been known to clamber out of the water in some very strange places. Imagine the surprise of a couple of kayakers in the Forth Estuary in Scotland in 2017, when not only did the whiskery Labrador-like face of a Grey Seal appear right beside them, but the animal clambered onto one of the boats. This has happened in several seal hotspots around the UK, and it seems it is more out of curiosity then necessity as there are often rocks nearby. However, in areas where seals are not used to boats and people, kayakers can cause disturbance by approaching haul-out sites too closely.

Other strange resting places used by Grey Seals include the swimming and diving platforms of moored boats, and navigation buoys, which are a favourite. In these cases, the seals are just making good use of what are essentially islands created by humans.

Below: This pair of Grey Seals has found a comfortable resting place on a starboard channel marker buoy.

Family life

In seal society it seems that the females do all the hard work of raising a family. Seal mothers are very attentive to their pups and will protect them at all costs. But family life is short, and the pups must grow up quickly and learn to survive on their own within just a few short weeks. The females have nothing more to do with their young once they are weaned and there is no extended family life. In contrast, many whale and dolphin species live and travel in family groups called pods. Male Grey Seals are more of a danger to small pups than anything else and can trample them when trying to mate with females ashore in the rookeries. The bulls wait for their chance to mate with as many females as possible, because the females become receptive once they have weaned their pups. The bulls find themselves a place on the shore where there are plenty of females and stay there, seeing off newly arrived males.

Below: In contrast to Grey, Common and other true seals (Phocidae), that remain on shore to suckle their young for days or weeks, sea lions may attend them for months, leaving to go on days-long fishing trips.

Above: Common Seals make attentive mothers and excellent role models for their young.

Common Seal males do not necessarily join the females on the breeding rookeries, but instead usually wait for them in the water. During the breeding season they patrol the water alongside the beach, fighting and jostling for a chance to mate. As with most of their habits, Common Seals usually mate while in the water. When the males need to haul out for a rest, they do so nearby, but mostly separate from the females.

Early swimmers

Common Seals give birth to a single pup, which in the British Isles takes place during summer (mostly in June and July). They do so in what might seem very precarious places, including intertidal and offshore sandbanks, but also choose quiet mud and sand shores in sheltered estuaries, or enticing beds of seaweed-covered rocks in the sea lochs around the rugged coastline of Scotland.

Common Seal pups are real water babies and can swim within a few hours of birth – which is just as well for those whose birth site is covered by the incoming tide. With only a single pup, the attentive mother can devote all her attention to it, and in the early days she keeps it close to her at all times, whether in or out of the water.

She will even clasp the pup between her fore flippers and submerge with it if danger threatens. Tired pups are sometimes given a rest on their mother's back.

Common Seal pups are rarely spotted suckling, but do so both in and out of the water. Females often suckle their pups right at the water's edge so they can dive underwater quickly if disturbed. Mothers have also been seen floating contentedly in the water with their pup gripping onto one teat with great determination. Wherever they are, the pups are suckled for three to four weeks. They spend a lot of time asleep at the haul-out sites, but some pup-tagging experiments have shown that even when they are only a few days old, pups can spend from a third to half their time in the water. As they get older and more mischievous and adventurous, the pups will wander into the water alone, with their attentive mothers following.

While the females are nursing their pups, they do not get much chance to hunt for fish and so lose weight, while the pups gain weight and put on fat rapidly. However, a month is a long time to go without food and towards the end, the mothers slip away for longer periods in search of a well-deserved meal.

Above: Grey Seals, like this one, suckle their pups on land whereas Common Seals will suckle both in and out of the water.

Foster mothers

It is enough work for a seal mother to feed and look after one pup, let alone two. Though twins are rare, both Grey and Common Seals are occasionally seen with two pups happily suckling together. Mothers will sometimes suckle an orphan, especially if she has lost her own pup, or a lost pup will get in a sneaky meal before the mother realises what is happening. Caring for two pups means the mother uses up more of her fat resources and the pups are less likely to gain as much weight before they are weaned. Most females will push away a strange pup, and it is probably the less experienced mothers that take on strays.

Born on the beach

In contrast to Common Seals, Grey Seals in the UK give birth in autumn and winter, with the exact timing depending on where they live. Between about September and December, expectant Grey Seal mothers haul out onto shores with space above the high-water mark for their pups. This is important because the pups can't survive for long in the water until they have moulted the warm white coat they are born with, which is not waterproof. The pups are usually born when the tide is going out, and as the water returns, they do their best to struggle up the beach to safety. Whenever possible, the females choose places with minimal disturbance both

Below: The fluffy white coat of a newborn Grey Seal is extremely warm, but is not fully waterproof.

Above: The contrast in colour between a Grey Seal mother and pup is obvious. Pups lose their white lanugo coat at their first moult, two or three weeks after they are born.

from people and predators. Outlying remote islands are a favourite – the west coast of Scotland provides many such places, as well as hundreds of hidden mainland coves, bays and sea caves. Here, the mothers suckle their young in relative peace. In storm-ridden Orkney and Shetland, Grey Seal pups sometimes crawl inland to avoid the crashing waves, and can be found in coastal fields along with the resident sheep.

The peace is shattered when the males come ashore in search of a mate or, usually, mates. Protective females with young pups give the males a hard time and chase them off, but they become receptive to the males as weaning time approaches. The mothers abandon their pups and go back to sea shortly after mating, but the pups usually stay on the beach for several more weeks, until they are brave enough and hungry enough to venture into the water. However, if there are suitable rock pools or sheltered areas nearby, some pups may learn to swim before they are weaned and have moulted. Sometimes, Grey Seal pups will swim when only a few days old, closely attended by their mother. Just like human children, some seem to love the water while others are initially reluctant swimmers.

In warmer areas such as the Isles of Scilly and Cornwall, Grey Seal pups can be spotted on shores in August and September. The milder climate and earlier onset of spring in the south means that they can cope with an entire winter. Further north, on the east coast of Scotland, most are born after Christmas in late December or January. The weather can be terrible here at this time of year, but it is only a few weeks before spring comes along. Even so, many pups born on shores backed by steep cliffs drown if a severe storm hits before they have learnt to swim. Like their Common Seal counterparts, Grey Seal mothers barely feed when suckling their young, which they do for three to four weeks.

Below: Finding Grey Seal pups in a coastal field would certainly be a surprise in most places, but in Orkney and Shetland the pups sometimes crawl inland to avoid storm-driven waves. This particular field is also a breeding site.

Icy homes

All seals have to come ashore to give birth to their pups, but what happens if there are no accessible beaches, especially if, like Grey Seals, the pups are born in winter? In the northern parts of their range, such as the Baltic and around Iceland and Norway, Grey Seals often have their pups on shore-fast ice (the ice that extends from the shore) and even on floating pack ice. These northern populations breed later than those in the British Isles, allowing the pups to be born after the ice has formed.

Left: Grey Seal pups born on ice in northern areas blend in well with their snowy home.

Below: With the winter melt coming ever earlier, Grey Seal pups born on ice around the shores of the Baltic Sea often have to retreat to less secure sites on the shoreline.

Lost coats

In order to keep their precious warm coats in good condition, all seals moult once a year, gradually losing their old, battered hair and growing strong new hair beneath it. At this time, Grey and Common Seals lose their sleek appearance and their individual spotty patterns, and sport tufts of tatty brown hair.

Common Seals begin to moult after the pupping season in the late summer, but the exact timing varies depending on where they live. In Scotland, moulting usually begins at the end of July or in early August and takes four to six weeks. Moulting can be a chilly experience, and so Common Seals stay hauled out of the cold water for much longer than usual. The new hair grows quickly because blood flow to the skin increases at this time. However, extra heat is consequently lost whenever the animal goes back into the sea, which is a very good reason for noisy humans and dogs to keep away from haul-out sites at this time.

Below: Two to three weeks after it is born, a Grey Seal pup loses its fluffy white lanugo coat and is then ready to go to sea.

Above: Moulting seals have a tatty and dishevelled look, but once the old fur has all been lost, they soon resume their sleek appearance.

Grey Seals moult in spring between March and May, about two months after giving birth. Breeding and the annual moult are the times when the largest numbers can be seen hauled out on the shore, sometimes hundreds or even thousands of animals. Many of the sites where they gather are traditional and are used every year. At moult time, Grey Seals spend even longer out of the water than Common Seals do, but they seem to go back into the sea at least once a day.

Baby Grey Seal pups are born as fluffy white bundles and cannot survive in the water until they have moulted and grown their waterproof adult coat. This first fur coat is called a lanugo coat. Common Seal pups are much the same colour as the adults, because they moult their white lanugo coat before they are even born. They need to do this because they are out swimming with their mother just a few hours after birth. If you spot a white seal pup on a beach around our shores, it will therefore be a Grey Seal and not a Common Seal – although some of our occasional visitors, such as Ringed Seals, also have white pups.

Speech and drama

Seals are mostly silent underwater, but on land they use their voices to good effect. Grey Seal rookeries can be noisy places, especially where the sound bounces back from a cliff-lined shore. Most of the noises seals make are either aggressive or for communication between mothers and pups. Females with pups howl and snarl to keep other adults out of their space, and mothers also make contact calls to help their pups find them if they stray along the beach. Just like a human baby, a Grey Seal pup will wail when it is hungry, and it soon learns to make 'keep away' noises when threatened. Males make aggressive, low-frequency roars and hisses to intimidate other males during the mating season.

Common Seals produce fewer sounds than Grey Seals, and in general they are fairly quiet when hauled out. They dislike lying too close to one another, and produce a hissing growl and slap their fore flippers to maintain their personal space. However, Common Seals do seem to use their voice underwater more than Grey Seals. During the breeding season, males make roaring sounds underwater while patrolling off haul-out sites, both to establish breeding territories and attract passing females.

Below: Although Grey Seal pups may look helpless, they have sharp teeth and soon learn to make warning sounds to see off unwelcome visitors.

Above: Mothers and pups use voice, smell and touch to find and communicate with one another. Females have a particularly good sense of smell for pup recognition.

Common Seal pups left on their own on the shore while their mother is away fishing make short, sharp calls, almost like a crow cawing. The females can recognise their own pup's call even if there are many others around. It might appear to us that a mewing, lonely seal pup has been abandoned, but Common Seals leave their pups alone from a very early age, returning periodically to feed them.

Common Seals have good hearing both in and out of the water, which is not the case with all seals. Some species hear much better underwater than on land, which makes sense because this is where predatory Orcas and sharks live. Common Seals and other species that haul out on ice floes also need to be alert to predators such as Polar Bears when out of the water.

Singing seals

From a distance, a seal colony in full voice can sound rather like an out-of-tune choir. It turns out that Grey Seals can actually sing – after a fashion anyway. Individual captive seals at the University of St Andrews have been taught to copy simple melodies, including 'Twinkle, Twinkle, Little Star'. This is not true singing, of course, but in simple terms the seals were able to copy the peak frequencies in the song, albeit after considerable training. The research showed just how flexible seal vocalisations are. Mothers and pups can certainly recognise one another's calls, even in the noisy throng of a breeding colony.

Life at Sea

Seals are excellent swimmers and are totally at home in water. However, like us and the other 5,000 or so known mammal species, they breathe air. As a group, mammals evolved on land, and only a few are genuinely aquatic, including 33 pinniped species. In the water, terrestrial mammals struggle to hold their breath, to stop themselves sinking and to keep warm, even in tropical seas, and they would undoubtedly drown if they tried to sleep in water. Seals have had to overcome all these problems.

True seals, including Grey and Common Seals, swim using their rear flippers. If you watch a seal swimming underwater, it looks as if it is sweeping a large tail from side to side, but what you are seeing is a pair of large, backwards-trailing, webbed flippers. A seal swims a bit like a salmon, its whole body bending sinuously from side to side as the flippers push against the water. The short front flippers are used almost entirely for steering, and seals also have a small pointed tail between their rear flippers that they sometimes use like a ship's rudder.

Not all pinnipeds swim like this, however. If you look at a sea lion when it is sitting up, you will see that its front flippers are much longer and stronger than the rear ones – the complete opposite of a Grey or Common Seal. If you watch one swimming, perhaps through an underwater observation window at a zoo, you will see that it does so in quite a different way to seals, moving its front flippers in and out from its body, pushing water backwards and propelling itself forwards.

Above: A sea lion's long front flippers push back against the water to provide forward thrust. The shorter rear flippers help with steering.

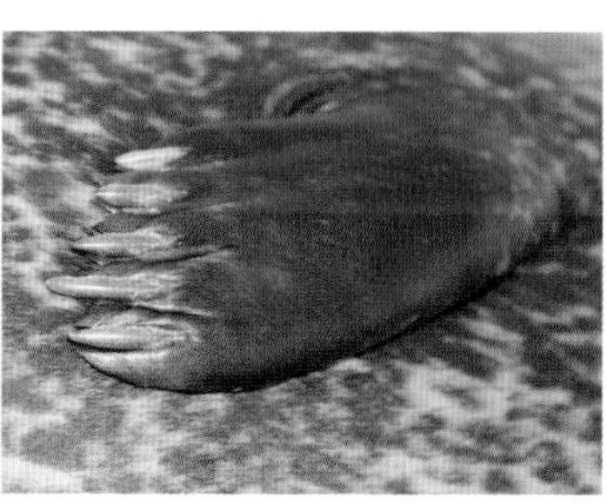

Left: Seals have five bony digits in each flipper, just as we do in our hands and feet. The front flippers are short, with equal-length digits, and the long, powerful rear flippers have two outer digits that are longer and wider than the three inner ones.

Opposite: True seals use strong sideways sweeps of their hind flippers to power through the water.

Diving deep

Most people can hold their breath and swim underwater for only one to two minutes, and perhaps up to five or six minutes with training (although there have been a few exceptional longer records). We simply do not have the capacity to store enough oxygen for longer dives, nor can we ignore the build-up of waste carbon dioxide in our bodies that forces us to breathe in. Seals do much better, and Southern Elephant Seals are record-holders in this respect, able to stay underwater for an astonishing two hours. Like most seals, Grey and Common Seals spend most of their time (about 80 per cent) submerged, hunting, playing and generally looking around. Grey Seals normally need to take a breath every four to eight minutes depending on what they have been doing, and they dive down again after less than a minute on the surface. For the smaller Common Seals, three minutes is a good dive. Exceptionally, both Grey and Common Seals can stay underwater for 30 minutes, but they must then spend longer at the surface to allow their body to get rid of all the waste products of respiration (breathing) that have built up and to restock their oxygen supplies.

Below: Strong muscles allow the Grey Seal to close its nostrils when it dives into the water. The only way humans can do this is by wearing a nose clip or pinching our nostrils shut with our fingers.

The question of just how deep each species of seal can dive is a difficult one to answer definitively, but as far as they are concerned they dive only as deep as they need to in their search for food. Although Grey Seals prefer to hunt near the seabed, they will catch fish on the way down if they can. On the other hand, if they know of a fish-rich rocky habitat in deep water, then they might make an effort to visit it. If necessary, Grey and Common Seals can dive down 300–500m (1,000–1,600ft), but most dives are shallower than 40m (130ft) and often much less.

Onboard oxygen

When a seal dives down, it can store extra oxygen in its muscles compared to land mammals. The oxygen is attached to a muscle protein called myoglobin, which is one of the reasons why seal meat is such a dark red colour. During a dive, the stored oxygen can be released into the bloodstream and carried around the body if supplies run low. Seals also have much more blood relative to their size compared to land mammals, including humans. As in all mammals, oxygen is transported around a seal's body by the blood, so extra blood allows an increased amount of oxygen to be carried.

Below: Seals can open and close their nostrils underwater and seem to blow bubbles just for fun, as well as doing it as a warning to other seals (or divers) to keep out of the way.

Above: Seals often breathe out before they dive so that they don't have to fight the buoyancy of a lungful of air as they swim down.

As soon as a seal surfaces after a dive, its heart rate goes up dramatically and blood is pumped quickly through the lungs. This means the seal can get rid of waste carbon dioxide and take in and store more oxygen in only a short time. A seal needs only a minute or so to recover after a normal dive, and can bob up and down all day without getting exhausted.

For longer dives, seals have another oxygen-saving trick – one that is also used by other diving mammals. First, their heart rate slows right down to just a few beats per minute, then the flow of blood to distant parts of their body – such as the skin, swimming muscles and organs like the stomach – is greatly reduced. In this way, blood flows mainly to vital areas, including the brain and the heart itself.

Together, these and other adaptations are called the mammalian diving reflex, or mammalian diving response, and happen automatically. The reflex is strongest in marine mammals, but it seems to happen to some extent even in land mammals, including humans when we go right underwater. New research now suggests that at least some seals may be able to trigger this diving response just before they actually dive down.

The bends

Human scuba divers can suffer from a damaging condition called decompression sickness, or 'the bends', which is caused by breathing compressed air underwater and coming back to the surface too quickly.

The air in a diving tank is compressed as it is forced into the tank, so is at high pressure. As divers descend in water, the pressure around them increases. They breathe the air in their tank through a diving regulator, which reduces the air pressure to that of the surrounding water. However, because water pressure increases with depth, the air divers breathe at depth is always at a higher pressure than air at the surface. Gas volume decreases as the pressure increases, so more air is inhaled with each single breath taken at depth than at the surface. The vital oxygen component of the tank air (about 21 per cent of the total) is used up by the diver's body, but the nitrogen component (about 78 per cent; the rest is rare gases) stays dissolved in the blood. If the diver ascends slowly to the surface, the nitrogen is released slowly back into the lungs and breathed out. But if the diver surfaces too quickly, the sudden reduction in water pressure means that the nitrogen bubbles out when it is still in the diver's blood. Much the same thing happens when you take the top off a bottle of fizzy drink and bubbles appear in the bottle. In the case of a diver, it can be very dangerous if the bubbles get into the brain or other vital organs.

Seals (and other diving mammals) do not take compressed air down with them when they dive, but any air in their lungs will be compressed as they descend and could be forced into their blood. This does not usually happen because, unlike us, they do not take a deep breath before diving and may even breathe out, relying instead on the oxygen stored in their blood and muscles. Seal lungs are structured such that the alveoli (the tiny sac-like structures from where oxygen and nitrogen are absorbed into the bloodstream) can collapse, pushing any gas into parts of the lung where it cannot be absorbed.

Above: Seals suffer few of the problems and none of the discomfort that humans do when diving at depth.

Keeping warm

Seals have two ways of keeping warm in the cold waters in which most of them live: by having a wonderful fur coat, and by having a thick layer of blubber. Grey and Common Seals have a blubber layer that is up to about 6cm (2½in) thick, whereas nearly half the body weight of male elephant seals may be blubber. The blubber layer is not always the same thickness throughout the year – female Grey Seals, for instance, lose a lot of this fat reserve when they are feeding their pups on shore. Presumably they not only get hungry at this time but a bit chilly too.

In cold weather you will see land mammals raising their fur, fluffing it up to trap more insulating air – much as birds do with their feathers. Even we get goosebumps, as our wispy body hair stands up in response to the cold. In contrast, seals cannot do this, and their fur stays flattened and streamlined when they are in the water. However, their fur is waterproof and keeps them dry and therefore warm. Anyone who has been soaked through

Below: Seals groom their fur using the claws on both front and rear flippers. Doing this helps to keep it in tiptop, waterproof condition.

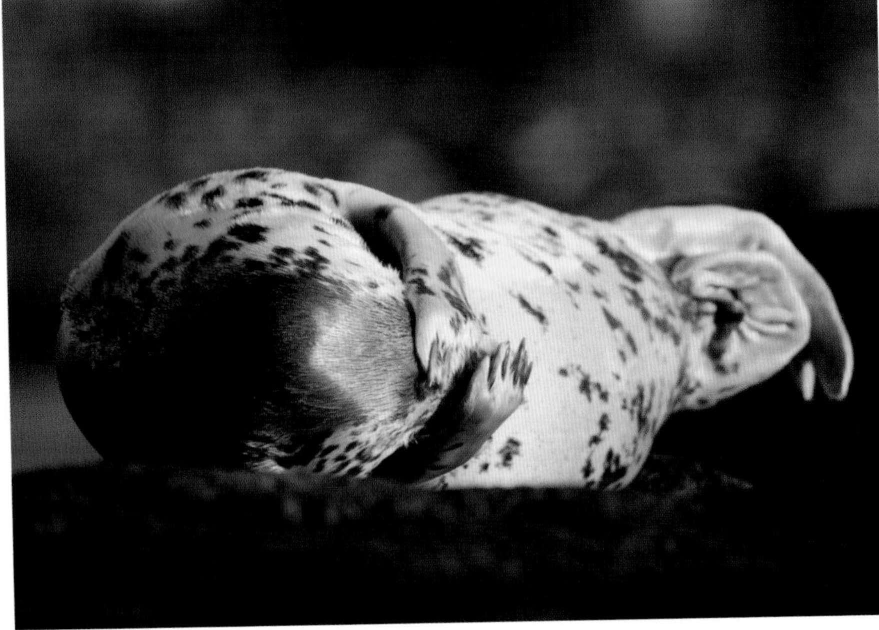

in a rainstorm will know just how much colder that feels than wearing the same clothes but with a good set of waterproofs over the top!

While we wash our hair regularly to stop it being greasy, seals continually secrete an oily substance from each hair follicle to keep their fur waterproof. The hairs are arranged in tiny tufts, with several fine hairs and a single longer and thicker 'guard' hair emerging from one opening in the skin. The fine hairs form a dense underfur and are protected from damage by the more deeply rooted guard hairs.

Above: Holding its large rear flippers aloft with the webbing splayed out helps this Grey Seal to cool down when hauled out on a hot summer's day.

Countercurrent

Seals, including our own Grey and Common Seals, have another trick up their sleeves (or flippers) to keep them warm. They have a heat-exchange system built into their flippers called the rete mirabile, a Latin term that means 'wonderful net'. The downside of having large flippers for swimming is that a lot of heat can be lost through them. To counteract this, hidden inside the flippers is a complex, intertwined network of arteries and veins that lie close to one another. Cold blood returning towards the heart from the flippers passes very close to arteries carrying warm, oxygenated blood from the heart and lungs in the opposite direction. Thanks to the rete mirabile, this warmth passes from the arteries into the veins and so is conserved.

In hot water

One of the disadvantages for seals of being covered in warm fur and thick blubber is that they can overheat in very hot weather, especially when they are hauled out on rocks and sand. Of course, diving back into the cool water helps, but that uses up precious energy. Shady rocks and caves and wet seaweed can also relieve the heat, and seals lying on sand can dig down to reach the cooler layers beneath. Grey Seals living in the warmer south of England often use another trick, splaying out their rear flippers and exposing the webbing between to increase heat loss.

Cape Fur Seals (*Arctocephalus pusillus pusillus*) living around South Africa have especially thick fur. In warm summer weather when lounging at the water's surface, they hold their large, wet flippers up in the air, using evaporation to help cool themselves off.

Below: Seals of all sorts, including these female Northern Elephant Seals on the central coast of California, will dig into the cooler sand beneath the surface on hot days.

Sleeping

Seals can obviously sleep and doze quite happily when they are hauled out on land, but how do they manage it when they are out hunting at sea? Both Grey and Common Seals can sleep for short periods by 'bottling' – hanging vertically in the water like an air-filled bottle, with their eyes closed and their snout poking up above the water's surface. This is easy in calm water, but must be difficult or perhaps impossible in wild seas. Seals also sometimes bottle when they are wide awake, so that they can look around and satisfy their natural curiosity. 'Logging' is another way of resting, in which the seal lies horizontally on its side.

Above: A Common Seal having a pleasant doze among Thong Weed (*Himanthalia elongata*) off the island of Islay in Scotland. It is obvious from this photograph why this behaviour is called 'bottling'.

Opposite: Seals are not the only marine mammals that can sleep vertically in the water. Sperm Whales (*Physeter macrocephalus*) do so underwater and at the surface.

Below: The sand dunes and flats at Donna Nook in Lincolnshire provide a safe haul-out site for Grey Seals to rest and to give birth.

Whales and dolphins can sleep with just half of their brain shut down and the other half still active, so can sleep underwater and remain alert to predators (although some also sleep at the surface). Experiments with captive fur seals have shown that at least one species can also do this, and that these animals did so most of the time they were sleeping in water. On land, some seals sleep using their whole brain and some with only half, and unlike cetaceans, they can choose which to do. Whether our Common and Grey Seals do the same is currently unknown.

Finding the way

When we open our eyes underwater, everything seems blurry. This is because our eyes are adapted to see in air, so if we want to see clearly underwater we must trap a layer of air in front of our eyes – by using goggles or a facemask. For seals, the opposite is true – they can see well underwater and have large eyes to cope with the dim light found at depth. On land, however, they cannot see very clearly, although they are sensitive to movement and have an excellent sense of smell.

Seals can find their way around and catch fish in poor visibility, when the water is full of silt or plankton. This is all down to their amazingly sensitive whiskers, called vibrissae. These long hairs sprout from two pads on either side of a seal's nose. Common and Grey Seals (and other true seals) also have small whiskery pads just above each

Below: Common Seals have one vibrissa above each nostril, three to five above each eye and about 44 on the snout.

eye and one or two whiskers above each nostril. These vibrissae allow a seal to picture objects clearly just by touching them. In fact, the vibrissae do not even need to touch the object – they can sense the shape, size and direction of the object just from the trail of water movement it leaves behind it (called the hydrodynamic trail). So a seal can follow the wake of a passing fish and catch it for dinner, or find a companion seal or a pup in the dark.

If we get especially cold in or out of the water, we lose our sensitivity to touch. Luckily, seals can move and use their vibrissae just as well in icy water as in warm water. This is because the follicles (hair 'roots') from which the vibrissae sprout are large and have an extra supply of blood to keep them warm, and the vibrissal pads contain a special type of fat that stays flexible even when it is cold (think of the difference between butter and soft margarine).

Below: Walruses are the most whiskery of all pinnipeds. They have around 700 vibrissae in their 'moustache', which they use to feel for food in the mud.

The Daily Routine

Whether land mammals live in rainforests, deserts or even urban environments, they need to eat, drink and avoid predators eating them. Although they bridge the divide between land and sea, seals' needs are no different. The unique ways that they breathe, dive, keep warm and sleep while in the water have already been described. But just as humans differ in their behaviour and preferences, so do individual seals.

Ocean hunters

Above: Grey Seals, like the above, and Common Seals have simple, mostly cone-shaped teeth that are good for catching and holding slippery fish. All seal skulls have large eye sockets to accommodate large eyes adapted for hunting.

Seals get all their food from the sea, which defines them as true marine mammals, and are primarily fish eaters. Adult Common Seals will catch and eat 3–5kg (7–11lb) of fish a day, while the larger Grey Seals need to eat around 5–7kg (7–15lb) a day. The amount of food a seal needs to eat depends on its size and also on the oil content of the prey. A 4kg (9lb) feast of sandeels (Ammodytidae; see page 69), for example, provides as much energy as a 7kg (15lb) serving of Atlantic Cod (*Gadus morhua*). Grey Seals hunt mainly for fish that live on and near the seabed, and will search out flatfish such as European Plaice (*Pleuronectes platessa*), European Flounder (*Platichthys flesus*) and Common Dab (*Limanda limanda*). With their sensory whiskers and acute sense of smell, the seals can track their prey down no matter how well camouflaged it is. White fish such as Atlantic Cod, Haddock (*Melanogrammus aeglefinus*), Ling (*Molva molva*) and Whiting (*Merlangius merlangus*) all swim quite close to the seabed and are also targeted by Grey Seals. Common Seals eat many of the same fish, but are more adept at catching them in mid-water, and will also target Atlantic Herring (*Clupea harengus*), Atlantic Mackerel (*Scomber scombrus*) and squid.

Opposite: This Grey Seal in Howth harbour, the Republic of Ireland, is enjoying a meal of Nursehound (*Scyliorhinus stellaris*) – a type of small, bottom-living catshark.

Above: Clasping a slippery Atlantic Mackerel with its sharp claws, a Common Seal takes a bite with its powerful, pointed teeth.

Thanks to air and sea transport, we can eat almost anything at any time of year, but exactly what a seal can catch depends on the seasons and which fish species are common in its home area. Like us feasting on locally grown strawberries, some seals might eat nothing but sandeels in summer and then switch back to a more varied diet as winter approaches and the glut is over. As for table manners, seals swallow small fish whole and can do so underwater. If a seal catches a fish that is too large to swallow, it will bring it to the surface and either shake it to pieces or hold it between its fore flippers and tear off chunks with its sharp teeth. However, like most other aquatic mammals, seals do not chew their food.

Slippery prey

In spite of their small size (20–30cm/8in long), sandeels are a very important food for Grey Seals. In some areas and in some seasons, more than half the diet of a Grey Seal can comprise sandeels, of which there are a range of different species. The flesh of sandeels is oily, providing plenty of calories for hungry seals. Sandeels are also favoured by seabirds such as Atlantic Puffins (*Fratercula arctica*), Razorbills (*Alca torda*) and Guillemots (*Uria aalge*). In the North Sea, many seabirds feed on almost nothing else during the breeding season and if sandeels are in short supply in any particular year, then the birds may not be able to breed successfully. Predatory fish such as Atlantic Cod also rely heavily on them. Sandeels feed on tiny zooplankton – mainly copepods and especially the species *Calanus finmarchicus*. As the climate changes and the oceans warm, the distribution of these tiny creatures alters and the sandeels follow. However, the seabird breeding colonies and the seals do not. Luckily, seals eat many different types of fish, but many seabirds rely almost solely on sandeels to feed their young.

Above: An Atlantic Puffin with a beakful of sandeels. Puffins have backwards-facing spines on the roof of the palate and a strong tongue to position fish with so that they can carry many at a time.

Below: A large shoal of sandeels swims in unison while feeding on plankton. If danger threatens, they can dive down into the sand.

Porpoise killers?

While Grey Seals are primarily fish eaters, there is now evidence that in some places, especially in the southern North Sea (including off the coasts of Belgium, France and the Netherlands), they may be preying on Harbour Porpoises (*Phocoena phocoena*). Even as adults these cetaceans are smaller than Grey Seals, although the majority of attacks seem to be on juveniles. Most of the evidence comes from photographs and from dead beached porpoises that have characteristic, though not always definitive, bite and claw marks. Grey Seals could simply be feeding on porpoises that are already dead, but this seems unlikely as many of the porpoises were healthy, well-fed juveniles at the time of death. More evidence comes from Grey Seal DNA found in bite marks on a few fresh carcasses. Perhaps most extraordinary is that this behaviour has now been seen and videoed from a boat off the coast of Pembrokeshire by ecologists working with Natural Resources Wales, a government conservation organisation. Grey Seals also attack and sometimes kill and eat smaller Common Seals, especially pups, behaviour that has been seen and photographed by members of the public.

Below: A fully mature Grey Seal is easily large enough to attack a small Harbour Porpoise like this one, but this is rare behaviour.

Left: Grey Seals have large mouths and a powerful bite and should always be treated with respect.

Feeding in the forest

In the UK, only about 10–15 per cent of land is covered by forests and woodlands and those that remain are no longer patrolled by large predatory mammals. However, off our rugged shores there is a different story to tell. Here, tranquil forests of large brown seaweeds known as kelp cover huge areas of the rocky seabed. Forest Kelp (*Laminaria hyperborea*) may not be as mighty as an oak tree, but it has a 'trunk' (called a stipe) that usually

Below: Both Grey and Common Seals use kelp forests as hunting grounds.

grows 1m (3ft) tall and, exceptionally, reaches 3.5m (11ft). This is topped by a wide blade (the frond), which is cut into long leaf-like fingers that spread out in the currents and waves to reach for the sunlight above. The shelter and food provided by these underwater forests attract fish of all sorts, making them favourite hunting grounds for Grey and Common Seals. These agile predators twist and turn their way between the kelp stipes as they hunt for flatfish and sandeels hidden in sandy glades, larger fish and squid swimming through the forest, and crustaceans and molluscs on the seabed.

Above: A male Cuckoo Wrasse (*Labrus mixtus*) would make a tasty meal for a seal.

Below: At low tide a forest of Oar Weed (*Laminaria digitata*) is exposed.

Further afield, Common Seals also hunt through forests of Giant Kelp (*Macrocystis pyrifera*) in the cool waters along the west coast of North America. This seaweed can grow to 40m (130ft) tall and provides similar rich pickings for California Sea Lions.

Drinking

'Water, water, every where, nor any drop to drink' bemoans the sailor in the famous *The Rime of the Ancient Mariner*. In this poem by Samuel Taylor Coleridge, the sailor knew that he would soon die if he gave in to the temptation of drinking seawater. So how do seals manage without fresh water? If fresh water is available, they may drink it, and some individuals have been seen eating snow, but there are very few records of this. Instead, seals extract sufficient water for their needs from their food and lose almost none of it through urine or by sweating.

Most seals mainly eat fish, which contain a lot of protein and fat and break down to produce water (called 'metabolic water') as a by-product of digestion. Every time a seal eats, it is bound to swallow some seawater, but its specialised kidneys can deal with this. The kidneys produce highly concentrated urine, containing more than twice as much salt as seawater and seven or eight times as much as their blood. Fish bodies have a similar salt content to that of a seal (and other mammals, including humans), and so a diet rich in fish does not add significantly to a seal's salt load. However, Crabeater Seals feed almost entirely on small shrimp-like animals called Antarctic Krill (*Euphausia superba*). These and other marine invertebrates such as crabs and shellfish are just as salty as seawater, so must add to the Crabeater Seal's salt load.

Below: A seal can close off its throat when opening its mouth underwater.

Natural enemies

Great White Sharks (*Carcharodon carcharias*) notoriously lie in wait for fur seals off South Africa's Western Cape province, shooting up and attacking them from below. However, these sharks do not (yet) occur in British waters and our seals are threatened by few predators – most of the UK's 21 resident shark species are far too small to be a problem for them. Other larger species are seasonal visitors, and at least 11 deepwater sharks live further offshore, but most of these eat mainly fish. Seal is on the menu for the huge Greenland Shark (*Somniosus microcephalus*) and Bluntnose Sixgill Shark (*Hexanchus griseus*), although Grey and Common Seals around our shores are unlikely to encounter these deepwater predators. Orcas are a different matter. They regularly visit UK coastal waters and there is at least one resident pod, off the west coast of Scotland. These intelligent animals are perfectly capable of catching and eating seals,

Below: Off the coast of South Africa, Great White Sharks make spectacular feeding attacks on fur seals.

including Grey and Common Seals, and do so regularly off Scotland's shores.

Baby Grey Seals spend several weeks on land before they learn to swim, and although they are safe there from sharks and Orcas, they face other dangers. While large gulls such as Great Black-backed Gulls (*Larus marinus*) are unlikely to attack healthy seal pups, they are great scavengers and are attracted to the afterbirth and dead pups left on the beach. A weak or sickly pup may also be pecked at, although there is little evidence for this. However, in the last 15 years or so Kelp Gulls (*L. dominicanus*) in Namibia have taken to attacking newborn Cape Fur Seal pups. They have learnt that if they peck a pup's

Above and below: Orcas around Scotland and in the rest of the North Atlantic mainly eat fish, but large ones sometimes kill seals.

Above: To a Black-headed Gull (*Chroicocephalus ridibundus*) the afterbirth from this newly born Grey Seal is a tasty treat.

eyeballs out, not only do they get a tasty treat, but the blinded pup is likely to die and can then be scavenged. Great Black-backs are one of the largest gulls in the world and like other gulls are clever and adaptable. Luckily, Grey Seal mothers are very protective, so hopefully our gulls will not adopt the behaviour of their Kelp Gull cousins.

Right: A juvenile Great Black-backed Gull eating a seal afterbirth. In doing this it is performing a useful clean-up service.

Lifespan

Captive Grey Seals can live for up to 40 years, but in the wild it is a different story. There, females may reach the ripe old age of 35, although 25–30 is more usual, while males are lucky to reach 20 years of age. Most males live only 10–15 years, worn out by weight loss and territorial disputes during the mating season. Female Grey Seals lose a lot of weight while suckling their young (see page 43), but bull seals can lose significantly more because they may be ashore for up to six weeks while waiting for each female in their patch to be ready to mate. In addition, they must continually fight off other males that try to take over their territory. When they do finally go back to sea, they must work hard to hunt and build up their blubber reserves again. Weakened and ageing bulls often die in winter, when they are in poor condition and the weather harsh. Common Seals can also live for 20–30 years, but much less is known about differences in the lifespan of males and females.

Below: A bull Grey Seal in prime condition, like this one, can be at least 2m (6.5ft) long and weigh up to 300kg (660lb), but may lose as much as half its weight during breeding.

Above: Grey Seal pups of various ages dot the marshy landscape at Donna Nook, Lincolnshire.

The first year of a seal pup's life is the most hazardous and a large proportion never make it to adulthood. Many Grey Seal pups drown when they are washed offshore into deep water during winter storms, or die when they get lost and starve to death. Common Seal pups can swim almost as soon as they are born, but even they can be overwhelmed by rough seas or get separated from their mother and starve.

Right: While most Grey Seal pups left on their own in the Horsey colony on the Norfolk coast have not been abandoned, some do lose their mothers and eventually die.

Seal characters

The Farne Islands in Northumberland are famous for the large colony of Grey Seals that live and breed there. These resident and visiting seals have been watched, counted, drawn, photographed and tagged since at least the 1940s. In 1962, Grace Hickling wrote *Grey Seals and the Farne Islands*, in which she describes how she came to recognise some individuals from physical peculiarities, and by squirting dye marks onto pups and putting tags on their flippers. Over the years she spent many weeks and countless hours watching the seals and recording their everyday lives.

What Hickling discovered was that seals are very individualistic. Each one differs in which fish it prefers to eat and how it catches them, and in its favourite haul-out spots. In the same way, some are bullies while others (of the same size and sex) are cowards, and some pups are adventurous on the shore while others prefer to stay safely hidden away.

In those days of black-and-white photography, it was difficult to recognise individuals from their coat patterns alone. Digital colour cameras have changed all that, however, and photo identification can now be used at haul-out sites to follow the everyday lives of seals (see page 96).

Above: Grace Hickling on the Farne Islands.

Above: Grey Seals still pose for the camera on the rocks in the Farne Islands.

Watching Seals

Sleek and silent, seals can vanish underwater like wraiths, there one moment and gone the next. But these pinnipeds are curious and like to know what is going on in their watery domain, so if you put in the effort to visit the right place at the right time, you may spot a whiskery face watching you. That said, spotting a seal in the water can be difficult. So also explore isolated islands, paddle into sheltered coves or peer quietly over dunes and rocks during a coastal walk – you just might be rewarded with the sight of seals sunning themselves on the beach.

Top spots

There is a real chance of seeing a seal almost anywhere around the coastline of Britain and Ireland. This is especially true in undisturbed areas, where quiet, discreet observation with a pair of binoculars will prevent the ungainly and rapid exit of the seal into the sea. However, the best chance of seeing seals in any numbers is either to go on a dedicated boat trip or a supervised visit to a known haul-out site. Such controlled visits are less likely to cause disturbance, especially at breeding colonies, and reserve wardens, volunteers and knowledgeable boatmen are usually on hand to guide you and tell you more. Some managed reserves even have webcams for a closer view. Many coastal RSPB nature reserves are excellent places for seal spotting as well as birdwatching, as are similar nature reserves run by other organisations all around the UK. On the next three pages is a small selection of top seal spots to visit.

Above: A path and fence at Donna Nook National Nature Reserve in Lincolnshire keep both seals and visitors safe, while allowing a close view of Grey Seal mothers and pups at the breeding colony.

Opposite: The seals at Blakeney Point in Norfolk are used to the regular, licensed boats and take little notice of them. This is a good way to get close to seals without disturbing them.

Monach Isles, Outer Hebrides

Landing on this set of five tiny uninhabited islands, a National Nature Reserve (NNR) west of North Uist, requires permission from Scottish Natural Heritage. But sailing or taking a boat trip from North or South Uist around this area is a joy, with the chance of seeing dolphins, Black Guillemots (*Cepphus grylle*) and Northern Fulmars (*Fulmarus glacialis*), which nest on sand dunes here instead of cliffs. Grey Seals breed here and the large colony produces many pups from October onwards. The Treshnish Isles in the Inner Hebrides are also important for breeding Grey Seals and can be accessed via boat trips from Mull.

Farne Islands, Northumberland

This group of 28 small rocky islands is famous for its history as well as its wealth of wildlife. The number and variety of islands make this area an ideal habitat for Grey Seals, and large numbers base themselves here throughout the year. The island group is owned and managed as a nature reserve by the National Trust. Boat trips run from Seahouses to Inner Farne, and at some times of year to Staple and Longstone. Dive boats visit a wider variety of sites during diving and snorkelling trips.

Donna Nook National Nature Reserve, Lincolnshire

In winter, Grey Seals come ashore along this 10km (6-mile) stretch of wild coastline to give birth to their white-coated pups. More than 2,000 pups were born at this NNR in 2018. The Lincolnshire Wildlife Trust operates a seal-viewing area here between late October and December, from where the seals can be observed in safety without disturbing them.

Blakeney Point, Norfolk

An extensive sandy peninsula curves its way along the coastline seaward of the village of Blakeney. This area is home to one of the largest Grey Seal colonies in England, much of it lying within a NNR. Groups of both Grey and Common Seals haul out onto the sandy banks here year-round. Locally operated boat trips run from nearby Morston Quay (mostly between April and October). These provide close-up views without disturbing the seals, who appear to recognise that the boats are not a threat and hardly lift a flipper at the many eyes and cameras peering at them. Nearly 3,400 pups were born here in 2019–20.

Skomer Island, Pembrokeshire

The only way to reach this 300ha (750-acre) island paradise is by boat from Martin's Haven, a small cove 14 miles south-west of Haverfordwest. A passenger boat runs on most days between April and September, excluding Mondays and weather permitting. October is the best time to see Grey Seal pups, but seals frequent the island year-round. Skomer also has spectacular breeding seabird colonies, while nearby Skokholm and Ramsey islands are also seal haunts. Skomer and Skokholm are NNRs, managed by The Wildlife Trust of South and West Wales, while Ramsey is an RSPB reserve with guided walks offered for a short period of the year.

Below: The official Skomer landing boat leaves from Martin's Haven on the mainland to take visitors to see the seals and birds for which Skomer Island (bottom) is famous.

Seals off course

Seals sometimes appear in unexpected places far inland. Unindustrialised estuaries make excellent hunting grounds for seals, and one or two will occasionally swim upriver in their search for prey. In January 2018, for example, a Common Seal was spotted in the River Great Ouse near St Ives in Cambridgeshire, almost 80km (50 miles) from the coast. Seals are spotted in this river in most years and also in the River Nene near Peterborough, having swum inland from The Wash. While most of these individuals are just adventurous or lost visitors, several Common Seals have stayed in rivers for at least two years – one even gave birth to a pup in 2014 and was photographed suckling it. The number of seals around the East Anglian coastline has been increasing steadily, and it may be that these predators are searching out new feeding and breeding opportunities away from the crowds at the coast.

Below: This heavily pregnant seal was spotted in 2010, having swum inland up a river in Cambridgeshire. She gave birth a few days after the photograph was taken.

Other wildlife

Many of the remote and beautiful shores where our seals rest and give birth are also home to a wide variety of other coastal wildlife. Seabirds and waders nest and feed along the ocean margins, and wildfowl graze the fringing saltmarshes. Coastal sand dunes and shingle spits blossom with salt-tolerant flowers in the summer months, while brown, red and green seaweeds sprout along rocky shores, and deep pools here shelter small fish, crabs, prawns and anemones. Further out, the sea is a playground for dolphins and the Harbour Porpoise, and you might even spot a whale or a Basking Shark (*Cetorhinus maximus*).

Eurasian Otter

The western coast and islands of Scotland are home to coastal-living Eurasian Otters (*Lutra lutra*). With the sea as an easy escape route, they can often be seen hunting during the day. Watch for ripple trails near the rocks on a calm day, and a small, pointed, whiskery head and lithe body may appear.

Above: Two otters rest contentedly on seaweed-covered rocks in Scotland.

Atlantic Puffin

In spring, Atlantic Puffins return to land to nest in clifftop burrows and on grassy islands. Protected nature reserves such as the Farne Islands, Skomer Island, Lundy Island in the Bristol Channel, and the Isle of May in the Firth of Forth all support colonies of Atlantic Puffins, as well as seals.

Right: An Atlantic Puffin on vigil near its burrow.

Above: A Sandwich Tern colony at Blakeney Point, Norfolk.

Sandwich Tern

Undisturbed sand spits are ideal for nesting tern colonies. The Sandwich Terns (*Sterna sandvicensis*) at Blakeney Point, Norfolk, share their roost with noisy Black-headed Gulls, which may be why the seals prefer a quieter spot around the corner. Little Terns (*Sternula albifrons*) also nest here.

Yellow Horned Poppy

Seals frequently haul out onto shingle and sandflats exposed at low tide, but if you miss them there is always something else to see. Above the high-tide mark, shingle is often more stable, allowing specialist plants to grow. Yellow Horned Poppies (*Glaucium flavum*) bloom in June and then produce extraordinary seed capsules up to 30cm (12in) long.

Right: Yellow Horned Poppy growing in shingle along the Norfolk coast.

Diving with seals

While seals are usually shy and difficult to approach on land, it can be a different story underwater. Many seals are naturally curious, and that certainly applies to Grey Seals. To the delight of divers and snorkellers, seals will often spontaneously approach and interact with them, especially in popular diving spots. The seals become used to seeing these strange and ungainly visitors, with their fizzing garlands of air bubbles. However, while a nipped fin or mouthed depth gauge may be harmless, divers should remember that these are wild animals with large teeth and should always be respected. Their natural curiosity is one of the reasons why seals are so successful. By continuously exploring and investigating new sights and sounds, they can discover new feeding opportunities and places to hide if danger threatens. This is not always appreciated by fish farmers – to a seal, a netful of lazily swimming salmon must seem like an invitation to dine.

Below: In some popular diving spots, seals become habituated to divers and will seemingly pose for underwater photographers.

Some boat operators run special trips to spots where there is a good chance of swimming with seals, but most encounters happen spontaneously. Some marine nature reserves and other marine protected areas have their own resident populations of Grey Seals. For example, Lundy Island in the Bristol Channel currently has around 200 resident seals and is well known for diver and seal interactions, and the Farne Islands off the Northumberland coast are another hotspot.

Seal-watching hints

Whether you are watching seals from the land or underwater, there are a few sensible and important guidelines that will help you get the most from the experience and protect both you and the seals from harm. Well-known seal sites may have noticeboards or wardens that can provide information. While seals attack people only very rarely, it does happen, both in and out of the water. In the Southern Ocean, Leopard Seals have occasionally stalked people and on one occasion an individual even drowned a researcher. Nothing like this has happened in the UK, but there have been a few occasions where swimmers have been mouthed or aggressively bumped by Grey Seals. Below are some recommended guidelines for watching seals on land and underwater.

On land

- Watch from a distance, especially in the breeding season. Seals are often nervous on land, particularly when they have pups. A dog or person on a rookery (breeding site) beach can cause pandemonium among the seals and result in injured or abandoned pups. A heavily pregnant female tearing back into the sea over the rocks might also unwittingly injure her unborn calf or even lose it.

- Keep dogs on a lead or leave them at home – you will probably see more seals if you do so.

- Abandoned baby seals may not, in fact, be abandoned at all. Grey Seal pups are left on the beach after weaning when only two to three weeks old, and Common Seal pups may also be left behind while the mother forages. Lone pups are therefore best left that way unless they are obviously injured or sick, in which case contact one of the organisations listed on page 125.

Above: When taking photographs, it is important to do so at a distance, and not to approach too closely like the people above. Access permission may be required at nature reserve sites.

Underwater

- When in the water, you are in the seals' environment and it should be their decision to approach you, not the other way around. Move slowly and look all around – including behind you. If you respect the seals, then their curiosity will often overcome their fear of you.

- Do not try to touch seals or entice them with food or alluring objects. If you do so, they can learn bad and potentially dangerous habits – this has been demonstrated many times over in land animals.

- Taking underwater photographs is a great way to keep and share your memories, but avoid using a flash if possible. You will often get better shots anyway and will not risk scaring your subject away.

- Stressed seals will often bare their teeth or become noisier. This is your cue to move quietly away.

Above: Seals use their mouths and whiskers to explore unfamiliar objects, such as divers' fins. Such close encounters can be exciting, but divers should move gently away if the seal seems stressed or antagonistic.

Seal science and surveys

The conservation and management of any animal, plant or habitat can be effective only if it is based on knowledge of the subject's biology, ecology and distribution. With our seal species, this means knowing, among other things, which areas around the UK are important haul-out sites for resting, moulting and giving birth; whether the seals stay in one area or move between sites or even countries, and when they do so; how many seals there are in total; and how many pups are born and survive to maturity. Much more is known about Grey Seals than Common Seals, simply because Common Seals spend so little time on land.

In the UK, much of this type of research is carried out by the Sea Mammal Research Unit (SMRU), based at the University of St Andrews in Scotland. Large conservation charities such as the Wildlife Trusts, the RSPB and the Marine Conservation Society also provide vital statistics. However, local knowledge is equally important and smaller organisations, including the Cornwall Seal Group Research Trust, as well as individual wildlife watchers, all

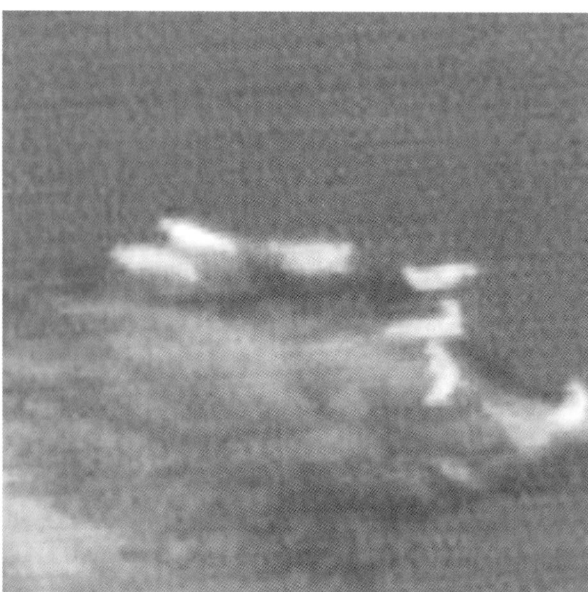

Right: Thermal images, such as this one are of lower resolution than optical images. That can make species identification more difficult, but the seals show up well, making it easier to count them.

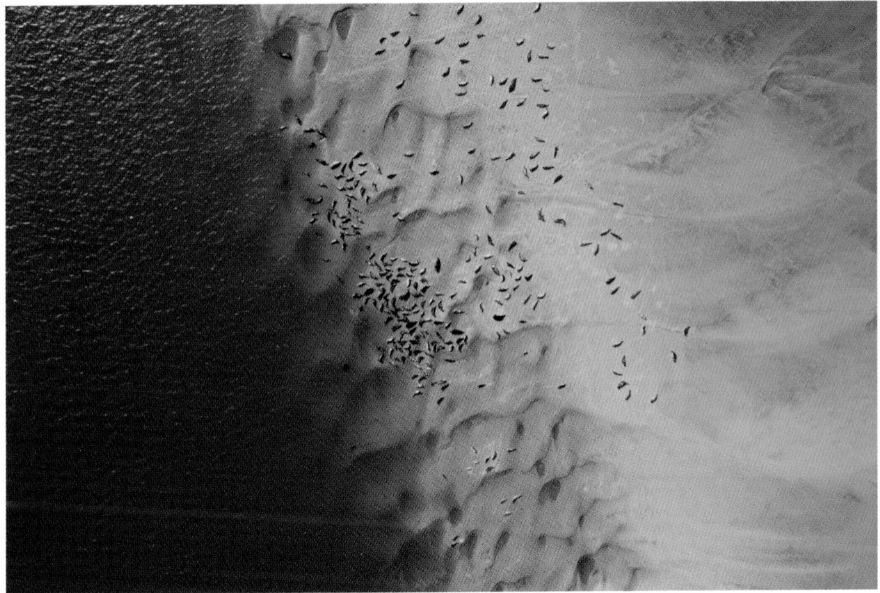

Above: Seals hauled out in sandy areas can be relatively easily counted using aerial photographs and surveys.

play their part. (See page 125 for website listings of these organisations.)

It takes a lot of patience and perseverance to watch and record data on seals. Sitting hunched up on a beach or in a boat in the pouring rain counting and observing seals through steamed-up binoculars takes dedication. Nowadays, technology makes things a little easier. Aerial photography and drones are used in seal counts, satellite tags and cameras fixed to individual seals record movements and migrations, and digital cameras and computers allow individual seal behaviour and movements to be tracked.

Counting seals

Each year, the SMRU counts the number of seals around the UK coastline. This is no easy task, because seals spend most of their time in and under the water. However, in the autumn Grey Seals come ashore to have their pups, which stay on shore for several weeks. The pups are counted at this time and the total population of Grey Seals is then extrapolated from these numbers using a mathematical model. It would be impossible

Above: Grey Seals lying around on rocky and seaweed-covered shores don't all show up well in photographs so thermal imaging cameras are often used to aid in counting them.

to check every single place where Grey Seals might be born, but the SMRU does count the pups in all the major breeding colonies. The seals are large enough to be seen from above, and the surveys are done mostly using aerial photographs taken from helicopters, light aircraft or drones. This method is especially useful in the remote and rugged landscapes of Scotland and Wales. In eastern England, the breeding colonies are relatively easy to reach by car, boat or on foot, and the pups there can be counted by trained volunteers as well as researchers.

Common Seal pups can swim as soon as they are born, so their numbers cannot be ascertained with any accuracy. Instead, Common Seals of all ages are counted when they come ashore to moult in August, because this is when they spend most time on shore. Many will still be in the water when the counts are done, but the surveys give the scientists a minimum estimate of their numbers. Only parts of the UK coastline can be covered each year, but the entirety is surveyed within five years or so.

Tagging seals

If you have ever wondered what happens to Grey Seal pups once they are weaned and their mother abandons them, you are not alone. Seal scientists ask this question too and have devised ways of finding out the answer. When the pups are on the beach it is relatively easy to catch them, and this allows researchers to clip a small plastic tag onto the membrane that stretches between their hind flipper toes. The pup will waddle away indignantly, but it is otherwise unharmed. When the seal dies, its tag can provide information on how old it was and where it was found in relation to its birth beach. Tags can sometimes also be seen and read on live seals.

While plastic flipper tags are cheap, the information they provide is limited. Modern telemetry tags are more expensive, but they are much more sophisticated, linking to satellites (or mobile phone networks) and sending back data for months at a time. They are used to find out where seals go on a daily basis to forage for food and how much time they spend in the sea and on land. Like a car satnav, telemetry tags use GPS to record the animal's

Below: A fin tag is clearly visible on the rear flipper of this unfortunate Grey Seal, pinned down by a larger male. Fin tags can be fitted quickly and easily in a similar way to the ear tags used on domestic cows and sheep.

physical position at any one time, and by measuring variables like temperature, pressure and light levels, they can also record how long the seal stays underwater and how deep it dives. The tags store this information and then transmit it to a satellite when the seal is at the surface or on land, and from there it is sent to scientists working on the project.

Below: Telemetry tags are glued to the head or back of an adult seal and usually fall off when it next moults. With luck, the tag can then be recovered and used again.

(c) Ryan Milne / S

Sound measurements

Seals have sensitive ears and can hear well underwater. They call and listen out for each other and for predators such as Orcas, and they may even be able to hear fish moving through the water. Oceans today can be noisy places, with boats and ships passing overhead, and construction works on jetties, oil rigs and wind turbines adding to the cacophony. In 2019, researchers fitted Grey and Common Seals in the North Sea with miniature sound and movement recording tags to find out what natural and man-made sounds they hear. The tags also measure pressure (and hence depth) and acceleration, take GPS readings, and record video to reveal what the seals are doing and how they react to noises.

Photographic identity parade

Common Seals come in many shades of grey and brown, but their coats are also beautifully patterned with small to medium-sized spots and, sometimes, pale ring-like markings and irregular blotches. Each individual has a unique pattern of spots, and with the use of high-resolution digital photography these markings can be used to identify them. By photographing all the seals at a particular haul-out site on numerous occasions, a local seal catalogue can be built up and individuals can subsequently be picked out in a photographic identity parade.

From this information, researchers may find that particular seals haul out at different sites on different days or at the same site over years, and patterns of movement can be worked out. Sighting histories of individuals, along with other data, are helping scientists in Scotland to work out why Common Seal numbers have declined drastically in parts of the north and east since the year 2000, but

Below: The most useful photographs for identifying individual seals are side shots of the head and neck.

Above: When photographing seals be careful not to get too close and disturb them. This seal is obviously not happy.

have stayed the same or even increased on the west coast. In Cornwall, the same technique has been used by the Cornwall Seal Group Research Trust to build up a photo identification catalogue of Grey Seals at important haul-out sites. Hundreds of seals have so far been documented.

If you live near the coast where seals haul out, you could try using this technique yourself – although take care not to disturb the animals. You will need a camera with a telephoto lens or zoom, a computer to store the photographs and a great deal of patience. There are as yet no national UK seal databases to which you can contribute your photographs, but local conservation organisations might well be able to use them or pass them on to researchers.

Messy science

Collecting faeces from beaches is one of the less glamorous jobs of a seal scientist, especially as it is often very smelly. However, this is a good way to find out just what seals have been eating. Their faeces contain undigestible remains such as bones, teeth and even skin,

Above: At the Cornish Seal Sanctuary in the village of Gweek, rescued seals are rehabilitated for later release. Visitors are welcome to watch and learn about seals.

which can be identified – researchers study owl pellets and otter spraints for the same reason. With so much plastic in the sea, scientists are now also using faeces to study whether seals are accumulating plastic fibres and fragments from the fish they eat – and it seems that they do. At Sea Life Trust's Cornish Seal Sanctuary in Gweek, researchers are working with the captive seals to find out if plastic is passing up the food chain. By analysing the wild fish that are fed to the seals and the resulting seal faeces, they can work out how much plastic is present. The scientists also extract DNA from the faeces for genetic studies of populations.

Unlike whales and dolphins, seals do not strand on beaches because they can live on land, but dead seals are sometimes washed ashore. Some of these have died of natural causes, including old age, but others may have died from starvation, a build-up of pollutants, disease, or injuries from boats and nets. To scientists and conservationists, a dead seal is a source of useful information, so if you find one on a beach, report it to one of the organisations listed on page 125.

Threats and Protection

Humans have hunted the seals around our shores for their skin, oil and meat from early times until the 20th century. As late as the early 1970s and 1980s, Common and Grey Seal pups were exploited for their skins in parts of the UK. Seals have also long been viewed as competitors for the fish we catch, and both our native species were subject to large-scale culls in the 1960s and 1970s. Today, attitudes and management methods have changed, and such gory massacres are hopefully a thing of the past.

Past hunting

Seals have always been important to indigenous people living in coastal communities around the world. Their tough waterproof skins have been made into cooking pots, fishing buoys and shelters, as well as shoes, belts, bags and clothing. Oil from their blubber kept lamps burning brightly, long before petroleum-based fuels were discovered in the mid-1850s, and their meat was an invaluable source of food. In Christian countries,

Opposite: A seal skin stretched over a frame to dry in Greenland.

Below: Native Alaskan Inupiaq hunters use a type of open canoe called an umiak, which is made out of seal skin stretched over a wooden or whalebone frame.

Above: Some remote coastal communities still rely on subsistence hunting. In this photograph, fresh seal skins dry at a camp on the island of Spitsbergen, part of the northern Norwegian archipelago of Svalbard.

where eating meat on Fridays was banned but fish were exempt, seals were also classed as non-meat and hence fit for the table.

Small northern and Arctic coastal communities took only what they needed to use and to trade with. However, others soon found that money could be made from seal oil and skins, and from the late 1700s thousands of seals were killed every year for profit, especially in Canada. In the early 1800s, imported barrels of oil rendered from Northern Elephant Seals were keeping the streets of London well lit.

As fur coats became fashionable in America and Europe, the white skins of juvenile Grey Seals living along the eastern coast of Canada were in high demand. When there were not enough left to hunt, Harp Seals became the next target. These ice-living seals were (and still are) abundant but much more difficult to reach, and the men catching them lived in terrible conditions. In 1982, the European Economic Community banned the import of baby seal skins, and in 1987 the Canadian government banned commercial hunting for 'whitecoats'. The clubbing, slaughter and skinning of such appealing animals for profit and fashion had become too much for animal welfare organisations and the general public worldwide to bear.

Seal culling

In the UK, both Grey and Common Seals were hunted until the early 1970s. However, commercial hunting was not so much the focus of concern as government-controlled culls. Culling involves reducing numbers for management purposes, and in the case of Grey Seals it was carried out for fisheries protection. Even as far back as 1958, a proposed cull of Farne Islands' seals hit the news headlines.

No licences have been issued for commercial seal hunting or large-scale culling in the UK since the late 1970s. A cull of Grey Seals was started in Scotland in 1977 by the then Department of Agriculture and Fisheries, but it was extremely unpopular with the public and environmental groups. The IUCN and the European Parliament decided that there was not enough evidence that seals were damaging fisheries and the cull was eventually called off after two years. Following this final cull, a considerable amount of effort was (and still is) put into gathering scientific data to show which fish species Grey Seals eat, in which areas, when and how much. To cut a long and very complex story short, the conclusion was that mass culling is an ineffective and extremely unpopular way of trying to make more fish available for commercial fishermen to catch.

Above: Seals may eat a lot of fish, but it is our intensive fishing methods that have caused some stocks of Atlantic Cod and other popular species to crash.

Left: Many fish stocks are managed and fished in a sustainable manner, which helps conserve fish for seals and people. These are Skrei Atlantic Cod fishing boats in Arctic Norway.

Legislation today

Since the early 1970s, when most exploitation stopped, regular scientific counts and estimates have been made of seal pup numbers born each year. These counts have shown that Common Seal populations in England have gradually increased overall, but with declines in 1988 and 2002 following major disease epidemics (see box on page 112). Grey Seal populations have increased more consistently over the decades, but the rate now seems to have slowed.

Today, seals in the UK are protected by various conservation laws, although this does not mean that they have complete protection everywhere and at all times. In England and Wales, seals are fully protected only during closed seasons (October to December for Grey Seals and June to August for Common Seals), although in some specific locations – such as the east and south coasts of England from Berwick-upon-Tweed south to Newhaven – protection is year-round. Seals are also protected year-round in Orkney, Shetland and the Outer Hebrides, on the east coast of mainland Scotland and in Northern Ireland. However, seals can be very destructive to some fisheries, and especially to aquaculture, and licences are issued allowing seals to be shot at times and in places where they are normally protected. Illegal shooting remains a problem in some areas.

Some of the UK's Special Areas of Conservation (SACs) have been chosen mainly because they are important places for seals to haul out or give birth. Nine SACs give protection for Common Seals and seven for Grey Seals. The well-known The Wash and North Norfolk Coast SAC is one, while the others are in the Highlands and Islands of Scotland.

Below: In 2010, the European Union banned the commercial import and sale of seal skins and other seal products. This seal-skin coat dates from the 1930s.

Ever-present threats

In spite of legal protection, seals – especially young ones – still face many problems and challenges caused by human exploitation of the oceans. We also eat fish, and catching them involves nets and ropes in which playful seals can easily become entangled. Many of the shores and coastal waters that seals use are littered with plastic, and while bays and estuaries provide seals with welcome shelter, levels of pollutants in these confined areas can be high.

Below: Without human intervention, this Common Seal will never get rid of its entangling trawl net. As the seal grows, the net gets tighter, causing painful and often lethal injuries.

Salmon dinners

In the UK we love salmon, which along with Atlantic Cod is a firm favourite. Most of that sold is farmed Atlantic Salmon (*Salmo salar*), and this delicious and nutritious fish is readily available from supermarkets and fishmongers. However, when seen from the point of view of a hungry seal, a fish farm is an invitation to dine. Inevitably, this leads to conflict between fish farmers and seals. In Scotland, where most of the UK's salmon farms are located, the farmers can legally shoot nuisance seals as long as they have a government licence; in 2017, for example, they shot 73 seals.

Above: Salmon fish farms vary in size from a few small cages to enormous installations. Any fish farm will inevitably be an attraction for seals.

Below: Acoustic deterrent devices can be used to keep seals out of harm's way during operations such as pile driving for new wind farms.

Wherever they are and whatever animals they raise, farmers need to protect their stock, but simply killing predators is increasingly being seen as unacceptable. As a result, salmon farmers in Scotland are now using non-lethal methods to deter seals. Seals are clever and can get through, over and under nets, but the new anti-predator nets that are being used on salmon farms do actually seem to keep them out. These nets have been fitted to several of the largest fish farms in Shetland, and salmon farmers there planned to have them on all their farms by the end of 2019.

In the Sound of Mull and other areas along the west coast of Scotland, many fish farmers are using acoustic deterrent devices (ADDs) to frighten seals away from their farms. ADDs produce a range of sounds that seals find thoroughly unpleasant. The problem is that whales and dolphins can also hear these noises from miles away. Although it has not yet been proved, there is some evidence that constant high noise levels may be driving cetaceans away from areas such as the Sound of Mull, where they were once common. New dolphin-friendly devices that aim to keep seals away but not cetaceans are now being tested. A seal startle device to stop seals taking the catch from nets set by fishing boats is also undergoing sea trials.

Ghosts in the sea

No fisherman wants to lose nets, but some inevitably end up drifting in open water or getting tangled over rocks and wrecks. Seals can easily become entangled in a net, and if it is heavy and the animal can't reach the surface, it will drown. Nets and ropes can also wind around a seal's neck or body, causing serious wounds that can then become infected. Drifting pieces of net can continue to catch fish and crustaceans, and when full their heavy load eventually drags them down to the seabed. The catch then rots or is eaten, and the lightened net rises again to fish another day. These 'ghost' nets are a particular danger to seals and dolphins, which are tempted in by the fish caught in them.

Below: Entangled with such a heavy load of rope and seaweed, this seal will be unable to hunt effectively and risks drowning if it tries.

Young seals are especially inquisitive and playful, and will investigate anything new and interesting. Floating plastic bags, plastic packing bands, tin cans and netting are all fascinating finds, and seals have been found with these items stuck around their neck and body. This is a particular problem for young seals, as the restricting adornments inflict deep wounds as the animals grow. On beaches backed by cliffs or in narrow inlets, slicks of plastic bottles and larger floating debris pose another threat for seals and young pups, battering the animals as the tide returns.

Seals are also caught in set nets fixed across estuaries and bays to catch wild salmon and other fish. The total by-catch from this and other fisheries includes an estimated 500 seals a year, most caught in the south-west of Britain, where these types of nets are more common. In these cases, the new startle devices being developed should help to keep at least some of the seals away.

Below: A carelessly discarded plastic bag is a plaything to a young seal until it gets stuck on its teeth or in its throat.

Disturbance

No one likes being disturbed when they are resting, but for seals repeated disturbance can be more serious. The sight, sound and smell of humans and dogs can all cause a seal to heave its way down to the water. Spooking heavily pregnant females or pups into a panicked rush over rugged, sharp rocks to the water is obviously not good for them, but even out of the breeding season seals need their rest. While they catch and eat their food in the sea, seals haul out on land to digest it. If they are disturbed, not only will it take them longer to digest their meal, but the extra activity uses up precious energy and warmth, especially in the moulting season, when the skin is flushed with blood. For these reasons, it is illegal to disturb seals at particular (designated) haul-out sites in Scotland, and in Northern Ireland it is an offence to disturb seals at any haul-out site intentionally.

Below: A disturbed seal will turn and look at you to assess the threat and is then likely to make a rapid exit into the sea.

Above: Wind turbines form a backdrop to the seal colony at Donna Nook, Lincolnshire.

Join the dots

Offshore wind farms are becoming an increasingly common sight around our shores, especially in the North Sea. Installing 80m (260ft)-high wind turbines is no easy task and it is certainly not a quiet job. Hammering the supports into the seabed certainly scares seals away, but records from tagged seals show that they soon return. Once the turbines are up and running, they produce a low level of underwater noise, but this does not seem to put seals off. Both Grey and Common Seals now appear to be visiting wind farms purposefully to hunt for fish. Not only that, but the SMRU has tracked seals and shown that some individuals swim in a grid-like pattern from turbine to turbine. The most likely explanation for this is that the seals have learnt that these 'artificial reefs' are good places to hunt for fish. Just as Red Foxes (*Vulpes vulpes*) have adapted to living in our cities, so seals may adapt to 'turbine cities'. What is certain is that with many more offshore wind farms planned, more seals will encounter them. But what is not yet known is whether the turbines are simply attracting and concentrating fish from surrounding areas, or actually increasing their numbers through better survival and reproduction.

Polluted waters

The appalling effects of spilt oil on seabirds is well known. Seabirds need to keep their feathers in good condition in order to stay waterproof and warm, and birds that are covered in oil swallow poisonous chemicals when they preen their feathers. Seals too are vulnerable to oil spills but rely mainly on their blubber to keep warm. However, that vital warmth-giving blubber can become a depository for other polluting chemicals.

Seals are top predators and so are the end store for persistent chemicals that can build up in food chains. While many pollutants break down relatively quickly in the ocean, some do not and continue to cause problems years or even decades later. Two of the worst offenders are dichlorodiphenyltrichloroethane and polychlorinated biphenyls, better known as DDT and PCBs, respectively.

Below: Seals often use a flipper to scratch at itchy skin and to groom, especially when they are moulting, but they do not usually use their teeth to groom their fur.

DDT is an insecticide and PCBs have a variety of industrial applications, and both were widely used for decades until they were banned in the UK and European Union in 1981. Sadly, the legacy of these notorious chemicals lives on, stored in seal blubber and passed on to the pups via their mother's milk. Research carried out at Abertay University in Scotland in 2018 has shown that these chemicals may be slowing down the rate at which Grey Seal pups put on fat. The significance of this is that a newly weaned seal that is thin is less likely than a fat one to survive long enough to learn to feed itself.

Below: Even on a remote shoreline in Namibia, in southern Africa, these resident Cape Fur Seals face disturbance and pollution threats.

Saving sick seals

If you find an obviously sick or injured seal (or dolphin or whale) anywhere in the UK, then the British Divers Marine Life Rescue (BDMLR) is a good organisation to contact for help (see page 125). It operates through a network of trained volunteers and has specialised rescue equipment. Although it is very stressful for a sick or injured seal to be chased and caught, these animals can recover well. Sometimes, all that is needed is to free the seal of entangling nets or other material. Others may need

Below: A seal will see a rescuer or researcher as a threat, so special training is needed to capture and care for wild seals.

a spell in a rehabilitation centre, such as the Hillswick Wildlife Sanctuary in Shetland, before being released. Note that rescuing seals can be dangerous to both you and the seal, and so should be left to the experts. If you are interested in becoming a BDMLR volunteer, the organisation runs regular training courses.

Prevention is obviously better than a cure, and some groups of trained recreational divers spend much of their time underwater collecting lost fishing nets and bringing them ashore. Nets are dangerous to divers as well as seals and other marine life, so this is doubly advantageous. Some groups have gone one step further and also arrange for the nets to be recycled into useful items.

Phocine distemper virus

Seals like to live in big groups, and as a result diseases can spread quickly through populations just as they do in humans. Phocine distemper virus (PDV) is a disease affecting seals that is related to the human measles virus and canine distemper virus. In 1988, an outbreak of PDV killed at least 18,000 Common Seals in northern Europe, especially around Denmark, Sweden and Germany, including 3,000 in the UK. Our worst-hit area was the Wash in eastern England, where more than half the seals died. In 2002, the disease hit again, but this time only about a quarter of The Wash population was lost and few seals died elsewhere in the UK.

No one knows for certain how the PDV outbreaks started, but it seems that both began near the Danish island of Anholt in the Kattegat. Small numbers of Grey Seals haul out at this site and mingle among large numbers of Common Seals, which is unusual. It is thought that these Greys could have carried the disease. Very few Grey Seals died during the epidemics, but this widely roving species can spread the infection to new areas. Harp Seals also carry PDV, and in the winter of 1987/88 large numbers of this Arctic species migrated south, probably from the Barents Sea in search of food (stocks of Capelin, (*Mallotus villosus*) an important part of the Harp Seal's diet, crashed that year). Whatever the ultimate cause, it is certain that some pollutants, including persistent PCBs, can lower immunity in seals, so both overfishing and pollution may have played their part. Will there be future outbreaks? Quite possibly, since blood samples have shown that immunity in our seal populations has declined since the 2002 outbreak.

Above: This picture, taken on 21 June 2007 on the island of Anholt in Kattegat, shows 27 dead seals that washed up on the beach. A mysterious virus killed the seals.

Seal of success

Compared to a century ago, the UK's seal populations are doing reasonably well. Grey Seals have a relatively restricted world range and our population represents about 34 per cent of the total. In contrast, Common Seals in the UK represent only about 5 per cent of the world population and the species is one of the most widespread among seals. The number of Grey Seals born each year in the UK has increased steadily since records began in the 1960s, but it now seems to be levelling off in most areas. Common Seals are not doing quite so well. In many parts of the UK, especially in Scotland, the Common Seal birth rate has been declining for some years, although this is by no means universal. Another outbreak of PDV could tip the balance in some areas.

Below: Like other young wild animals, newborn Grey Seal pups face many natural challenges such as diseases, parasites and bad weather.

Seals in Our Lives

Sleek, slippery and sinuous, seals have wound their way into people's lives for centuries. For some indigenous people, they remain a vital resource, providing food, clothes and an income. Fishermen might consider seals to be nothing but pests, tearing their nets and raiding salmon farms. For boatloads of tourists and divers, meanwhile, seals provide fascinating encounters, and for conservationists they can be ambassadors for their cause. For others, seals have a strong cultural and even spiritual significance.

Myths and stories

Appearing out of nowhere and disappearing just as quietly, seals are – not surprisingly – frequently the stuff of myths and legends. Perhaps the best known of these concern the selkies, who are mysterious seal people. The stories vary, but in all of them selkies are seals that shed their skin and change form to become human, at least for a while. If the skin is stolen while the selkie is in human form, he or she cannot return to the sea without it. There are many tales about lonely fishermen who find and hide female selkie skins so that they can marry these beautiful women.

Most selkie folklore originates in remote areas of Scotland, especially the Hebrides, Orkney and Shetland, and further north in the Faroe Islands and Iceland. These are all places where seals are especially common and have been a familiar sight to locals from the earliest times of human habitation. Like so many other myths, the origin of the selkie legend is unknown, although there are plenty of theories. Far back in time, what would our ancestors have thought if they saw someone in a remote seal-inhabited cove putting on a sealskin and then disappearing from view? Perhaps the figure was drying out his sealskin coat or the sealskin cover of his kayak. Or perhaps not.

Above: In 2007 a series of 10 Faroese postage stamps were produced, that depict the legend of Kópakonan or the Seal Woman.

Opposite: A bronze statue of Kópakonan was erected in 2014 on the Faroese island of Kalsoy. The statue is sometimes submerged by violent storm waves but survives, like the legend it celebrates.

Selkie stories live on today and are still celebrated. In 2007, an artistic series of Faroese postage stamps depicting the 'Seal Woman', or Kópakonan, was issued. This figure is also commemorated on Kalsoy, one of the Faroe Islands, where a beautiful bronze statue of Kópakonan was erected on the dramatic shoreline in 2014. In the UK, the Selkie sports brand is associated with open-water swimming and adventure. It's possible that the designers were inspired by the similarity between a wetsuit-clad swimmer and a seal.

In many selkie stories, the children born to selkie women and human fathers reveal their seal origins through restless behaviour. Many also have webbing between their toes and fingers. In real life, some people do have webbed toes or fingers, which is a harmless condition called syndactyly, where two or more digits are joined along part or all their length. It is easy to imagine how this could have added to the stories and legends.

Fictional pinnipeds

Seals are the heroes of many children's stories and films, where their endearing charm and ability to interact with people make them interesting subjects. Who wouldn't want to make friends and travel the fictional world with such sleek and whiskery companions? Many of the stories

Below: A fine-looking Walrus depicted in an illustration from Lewis Carroll's famous children's book *Through the Looking-Glass*.

Left: The Beatles album *Magical Mystery Tour*, released in 1967, includes the song 'I Am the Walrus', which was written by John Lennon and inspired by Lewis Carroll's poem 'The Walrus and the Carpenter'.

involve sea lions and fur seals, perhaps because they are the pinnipeds most often found in circuses and zoos. In *Finding Dory*, the 2016 computer-animated sequel to *Finding Nemo* (2003), Fluke, Rudder and Gerald are sea lions. The Walrus also features in stories, perhaps most famously in Lewis Carroll's nonsense poem 'The Walrus and the Carpenter', which appears in his 1871 book *Through the Looking-Glass*.

With their white-coated charm, Grey Seal pups go on fictional adventures – some based on truth – in many modern young children's books. Sandra Klaassen and Benedict Blathwayt both grew up or lived for a time in different parts of the Hebrides in Scotland and wrote, respectively, *Finn the Little Seal* (2019) and *Little Seal* (2017). Many Grey Seal pups are born around Hebridean islands, and presumably, the authors were familiar with their lives. Through these and other books, children who have never seen a live seal can also make the connection with the sea in these times of concern over marine conservation.

Neither Grey nor Common Seals seem to feature very much in the classics. English author Rudyard Kipling wrote many books about animals, collected together in his famous *Jungle Book* and published in 1894. One of his stories is 'The White Seal', featuring Kotick, a seal pup born on a remote island in the Bering Sea. After travels with his mother, Kotick returns to his home island and sees many pups being clubbed to death by the Aleutian islanders. The story relates how he leads thousands of seals to a safe place where there are no people. It is unclear exactly which species Kotick is, but he could be a Common Seal as the range of this species includes the Aleutian Islands. Common Seal pups are not white, so this would fit with Kotick being special – he is the only white seal in the story. Whichever species Kipling based his story on, this tale was an early recognition of the way humans slaughtered many northern seal pups for profit.

Below: Hugh Lofting's illustration of Doctor Dolittle releasing Sophie the seal into the sea.

"He threw Sophie into the Bristol Channel"

The seal in Hugh Lofting's *Doctor Dolittle's Circus* (1924) comes from Alaska and so could also be a Common Seal. Sophie is in a circus (so perhaps she is a sea lion) but is not happy. Eventually, Doctor Dolittle rescues her by dressing her up as a woman and throwing her over a cliff in Bristol (presumably not a high one). Unfortunately, he is then accused of murder. Again, the message is clear – the story is an early indication that not everyone was happy, even in the 1920s, with keeping seals and sea lions in captivity.

Art and sculpture

While it is not difficult to find a public statue or sculpture of a dolphin in UK towns and cities, pinnipeds hardly feature, and those that do are almost always sea lions. An amusing and aristocratic-looking sea lion is draped over the wall surrounding Cardiff Castle in Wales, along with a line of other animals (there was once a zoo here). In London, one of two posts of the Canada Gates leading into the Victoria Memorial in front of Buckingham Palace is topped by a 'seal', a netful of fish and a figurine. The seal has external ears, so is undoubtedly another sea lion or a fur seal. In Bristol, a wet sea lion plays in the fountain outside the Victoria Rooms. In grand buildings elsewhere, the occasional Walrus can be found decorating the tops of pillars. Our native seals do not seem to feature, however, perhaps because these statues

Left: The 'seal' depicted in London's Canada Gate statue is probably a Steller Sea Lion (*Eumetopias jubatus*) or a Northern Fur Seal (*Callorhinus ursinus*), since the gates were a present to Britain from Canada, where these two species are found.

Right: Nelson the Grey Seal lives on in Looe, Cornwall, in the form of a bronze statue, fashioned by the Cornish sculptress Suzie Marsh.

Below: Walrus tusk scrimshaw usually depicts the animals and people of the places where Walrus live. This early Inuit carver has included Polar Bear, reindeer and his contemporaries.

date from Victorian and Edwardian times when London and other city sculptors would only have seen pinnipeds kept captive in zoos.

Modern statues of seals usually commemorate individual seals that have made their home with or among people. In West Looe in Cornwall, for example, there is a statue of Nelson, a Grey Seal that lived for many years in the harbour here. He had only one eye and so was easy to recognise. Nelson died or disappeared in 2003, but he lives on in the form of a bronze statue on the harbour rocks and has become part of Looe's maritime heritage. At a much smaller scale, seals are very popular as ornaments and soft toys, although in most cases the species is not discernible.

Teeth and tusks of marine mammals have often been used by indigenous people, travelling sailors and whalers, and even prisoners as a basis for beautiful carvings. These decorative pieces of traditional art are called scrimshaw. Many are made from whalebone and Sperm Whale teeth. However, Walrus tusks were also popular – there are examples of Walrus scrimshaw dating from the 19th century in the British Museum.

Seal friends

Seals can become remarkably tame and seem to adapt well to human company. There are plenty of stories, both within and outside the UK, of wild Common Seals sharing their lives with people. Some are abandoned or injured pups that are adopted, but in some cases (such as Looe's Nelson) the seal befriends people. One of the best known of these friendships is that of British author Rowena Farre and a Common Seal called Lora, with whom Farre shared much of her younger life. Lora came into Farre's life when she was separated from her mother after a gale on the Isle of Lewis in the Outer Hebrides. Farre was on holiday there when a fisherman gave her the baby seal, which she took back to live with her in her aunt's remote croft in Sutherland. Farre relates this story in her bestselling autobiography, *Seal Morning*, published in 1957. Lora was a remarkable animal and seemed to adapt well to living, sleeping and playing in a small house,

Below: The truth behind many of the details of Rowena Farre's charming autobiography are contested by some. Nevertheless, young seals are curious and adaptable and certainly capable of learning new activities.

Above: The limestone statue of Andre the seal on the waterfront in Rockport, Maine, USA, is popular with tourists and residents. In 2018 cracks in the face were repaired.

just like a pet dog. She was free to swim in a freshwater lochan and had Eurasian Otters for company. Lora also supposedly learnt to play a toy trumpet, but this seems rather unlikely.

Another famous friendship, this time on the other side of the Atlantic in Rockport, Maine, also involved a Common Seal. In this case, the pup was purposefully captured, which back in 1961 was not illegal. The pup, called Andre, was raised with the Goodridge family and, once old enough, had the complete freedom of Rockport Harbour. He lived with the Goodridges for 25 years and became a tourist attraction, as well as a bit of a nuisance to the local fisherfolk and boat owners. Just like a dog, Andre fitted in with everything the family did, but he would also disappear for days or weeks at a time. He died in 1986 but is memorialised in the 1975 book *A Seal Called Andre* and in the form of an elegant statue at Rockport Harbour.

Seals are not the only marine mammals to befriend humans – dolphins are particularly renowned for this behaviour, and in the UK there have been several examples. These so-called 'ambassador' dolphins are invariably juvenile males who have either lost their pod (family group) or been pushed out of it for whatever reason. They search out places such as fishing harbours where they may get free discards from fishermen and can stay around for months or years before moving on. Ambassador Grey and Common Seals are also known – these animals choose to find and remain nearby people without being reared or kept by anyone.

Below: Bottlenose Dolphins are curious and often interact with divers. Some lone male dolphins seem to seek out human company.

Seals in captivity

Seals are very appealing animals, especially the pups, whose large, dewy eyes and often white coats attract widespread attention. Several species breed well in captivity, although it is now rare to see Grey or Common Seals in zoos or aquaria in the UK, or exhibited purely for entertainment. However, rescued sick and injured seals and abandoned pups are often taken to seal sanctuaries. Tourists are encouraged to visit these sanctuaries to learn about seals and provide support for conservation efforts (see page 125).

Below: While most baby seals cared for at the National Seal Sanctuary in Gweek, Cornwall, are released after rehabilitation, a few with particular problems remain as permanent residents.

Further Reading and Resources

Anderson, S. 1990. *Seals*. Whittet Books, London.
Written for the general public, this book about British seals is an easy and informative read.

Archer-Thomson, J. and Cremona, J. 2019. *Rocky Shores*. Bloomsbury, London.
An account of the marine life, communities, rocky places and landscapes around the coasts of the UK.

Blathwayt, B. 2017. *Little Seal*. Birlinn, Edinburgh.
Set in the Hebrides, this charming book is for young children.

Butterworth, C. and Nelms, K. 2014. *See What a Seal Can Do*. Walker Books, London.
A nature storybook about Grey Seals.

Dickenson, V. 2016. *Seal*. Reaktion Books, London.
A mainly cultural history of pinnipeds worldwide.

Dipper, F.A. and Pattulo, A. 2018. *Pocket Guide to Whales, Dolphins and Other Marine Mammals*. Lincoln Children's Books, London.
An attractively illustrated and fun book designed to help children learn about the wonderful world of marine mammals, including seals.

Farre, R. 1957. *Seal Morning*. Hutchinson & Co., London. Reprinted 2013, Birlinn, Edinburgh.
An autobiography describing the author's friendship with a Common Seal.

Goodridge, H. and Dietz, L. 1975. *A Seal Called Andre*. Down East Books, Rockport, ME.
The true story of a Common Seal living with a family in Maine, USA.

Hewer, H.R. 1974. *British Seals. New Naturalist Library 57*. HarperCollins, London.
A comprehensive and classic natural history of our native seals.

Hickling, G. 1962. *Grey Seals and the Farne Islands*. Routledge and Kegan Paul, London.
A fascinating early observational study.

Jarrett, B. and Shirihai, H. 2006. *Whales, Dolphins and Seals*. Bloomsbury, London.
A superbly illustrated guide to marine mammals for those wishing to learn more about species identification, behaviour and distribution.

Klaassen, S. 2019. *Finn the Little Seal*. Floris Books, Edinburgh.
Set in the Outer Hebrides, this tale of a Grey Seal pup is aimed at three- to six-year-olds.

Lambert, R.A. 2002. *The Grey Seal in Britain: A Twentieth Century History of a Nature Conservation Success*. Environment and History 8(4): 449–474.
A journal article on the changing reaction to seals over time.

Sayer, S. 2012. *Seal Secrets: Cornwall and the Isles of Scilly*. Alison Hodge Publishers, Penzance.
This small but fascinating book about Grey Seals is based on observations by the author, Sue Sayer.

Conservation groups

British Divers Marine Life Rescue (BDMLR)
bdmlr.org.uk
A charity with a network of volunteers trained to rescue marine mammals, including stranded cetaceans and injured seals. You can call on and use the service or volunteer for training.

Cornwall Seal Group Research Trust
cornwallsealgroup.co.uk
A conservation charity that collates records of seal sightings in south-west UK. You can become a citizen scientist and survey seals in your local patch.

Cornwall Wildlife Trust Marine Strandings Network
cornwallwildlifetrust.org.uk/strandings
The licensed recorder for dead stranded marine seals, cetaceans, basking sharks and turtles in Cornwall. You can report finds to them or train as a volunteer to photograph and record finds.

The Mammal Society
mammal.org.uk
A mammal conservation charity that aims to collect and share ecological and distributional information on UK mammals. You can submit records, join activities and use online resources.

The RSPB
rspb.org.uk
The largest wildlife conservation charity in the UK. You can join and use their resources, visit reserves and volunteer.

Scottish Marine Animal Stranding Scheme (SMASS)
strandings.org
Collates data on dead stranded seals, cetaceans, basking sharks and turtles around Scotland. You can report finds to them.

Sea Mammal Research Unit (SMRU)
smru.st-andrews.ac.uk
An organisation in the University of St Andrews that carries out innovative research on the biology of marine mammals. You can use their online resources to learn more about seals.

Wildlife Trusts: Marine Sightings and Strandings
wildlifetrusts.org/wildlife-advice/marine-sightings-strandings
There are 46 Wildlife Trusts in the UK, Isle of Man and Alderney. The website provides advice on what to do and where to report sightings and live and dead strandings of marine animals.

Seal sanctuaries

Cornish Seal Sanctuary
sealsanctuary.sealifetrust.org
A sanctuary run by the Sea Life Trust and located in the village of Gweek, Cornwall, that rescues and rehabilitates seal pups. You can visit the sanctuary, see the seals and join in activities.

Hillswick Wildlife Sanctuary
hillswickwildlifesanctuary.org
A sanctuary based in Shetland that cares for and rehabilitates sick, injured and abandoned seals and otters. Contact to ask about visits.

Seal Rescue Ireland (SRI)
Sealrescueireland.org
A sanctuary based in County Wexford that rescues sick, injured and abandoned seals from around the coast of Ireland, rehabilitates and releases back into the wild. There is a visitor centre and an education programme.

Acknowledgements

I was delighted to be offered the chance to write a book in this beautifully illustrated and produced RSPB Spotlight series. My thanks go to the whole team at Bloomsbury, especially Julie Bailey and Alice Ward, who made the whole process run so smoothly and to the picture researchers. Relying on my photographic abilities would not have been a good idea.

Sitting in the autumnal sunshine on a grassy clifftop in South Pembrokeshire recently, I could hear the plaintive cries of Grey Seal pups in a secluded cove far below and see through my binoculars their clumsy attempts to suckle from their sunshine-satiated mothers. I was walking part of the Pembrokeshire coastal path. The coasts around this area and much of the British Isles have been protected for wildlife and yet made accessible to us by many conservation organisations. I am grateful to them all, including the RSPB, National Trust, Wildlife Trusts, and the various government conservation agencies. Things were not so easy for early seal naturalist pioneers such as Grace Hickling, working during and just after the second world war. I found her pioneering study *Grey Seals and the Farne Islands* a mine of fascinating and detailed information.

Finally, my thanks to my husband, John Buckley, who patiently and consistently never believes me when I tell him this is the last book I'm going to write.

Image Credits

Bloomsbury Publishing would like to thank the following for providing photographs and for permission to reproduce copyright material. While every effort has been made to trace and acknowledge all copyright holders, we would like to apologise for any errors or omissions and invite readers to inform us so that corrections can be made in any future editions of the book.

Key t = top; l = left; r = right; tl = top left; tcl = top centre left; tc = top centre; tcr = top centre right; tr = top right; cl = centre left; c = centre; cr = centre right; b = bottom; bl = bottom left; bcl = bottom centre left; bc = bottom centre; bcr = bottom centre right; br = bottom right

AL = Alamy; FL= FLPA; G = Getty Images; NPL = Nature Picture Library; RSPB Images = RS; SS = Shutterstock, iStock = iS

Front cover t Tim Hunt/RS, **b** Arterra/ Contributor/G; **spine** Tim Hunt/RS; **back cover t** Steve Knell/RS, **b** Genevieve Leaper/RS; **1** SS; **3** SS; **4** SS; **5** SS; **6** t Bristol Museums, Galleries and Archives, b SS; **7** SS; **8** tl SS, tr SS, b SS; **9** t SS, b SS; **10** t SS, b Ingo Arndt/NPL; **11** SS; **12** SS; **14** t SS, b SS; **15** t Eric Baccega/NPL, b Fred Breummer/G; **16** SS; **17** Rod Teasdale; **18** SS; **19** SS; **20** Jim Brandenburg/NPL; **21** t SS, b SS; **22** t SS, b SS; **23** SS; **24** SS; **25** SS; **26** SS; **27** SS; **28** Stocktreck Images, Inc/AL; **29** Lillitve/G; **30** tl SS, tb SS, b Matt Cardy/ Stringer/G; **31** SS; **32** Erica Olsen/FL; **33** SS; **34** SS; **35** Erica Olsen/FL; **36** SS; **37** Rod Teasdale; **38** Education Images/ Contributor/G; **39** SS; **40** SS; **41** SS; **42** Suzi Eszterhas/FL; **43** Kevin Elsby/FL; **44** SS; **45** SS; **46** Peter Reynols/FL; **47** t SS, b SS; **48** Nick Upton/NPL; **49** Paul Sawer/FL; **50** ImageBroker/FL; **51** Suzi Eszterhas/FL; **52** SS; **53** t SS, bl & br Â © Biosphoto/Robin Monchatre/FL; **54** SS; **55** SS; **56** SS; **57** Huw Thomas/G; **58** Niall Bevie/NPL; **59** Arterra/ Contributor/G; **60** SS; **61** Laurie Campbell/ NPL; **62** KMGS Photography/G; **63** Kerstin Meyer/G; **64** Alex Mustard/NPL; **65** Paul Sounders/G; **66** SS; **67** Bristol Museums, Galleries and Archives; **68** Arterra/G; **69** t SS, b Graham Eaton/NPL; **70** SCOTLAND: The Big Picture/NPL; **71** t Alex Mustard/2020VISION/NPL, b SCOTLAND: The Big Picture/NPL; **72** t Alex Mustard/ NPL, b Robert Canis/FL; **73** Alex Mustard/ NPL; **74** SS; **75** t SS, b Michael Durham/FL; **76** t Jason Richardson/Al, b SS; **77** Niall Bevie/NPL; **78** t SS, b SS; **79** t © Tyne & Wear Archives & Museums / Bridgeman Images, b Nogel Eve/G; **80** Alan Spencer Norfolk/AL; **81** SS; **82** t Creative Commons CC0 License, b A.P.S. (UK)/AL; **83** t SS, b Anton Sorokin/AL; **84** SS; **85** Simon Stirrup/ AL; **86** t James Warwick/G, b SS; **87** t Gary K. Smith/NPL, b Frances Dipper; **88** Dan Kitwood/Staff/G; **89** t Les Gibbon/AL, b Alex Mustard/NPL; **90** Sea Mammal Research Unit; **91** Education Images/ Contributor/G; **92** Michele D'Amico supersky77/G; **93** Anterra Picture Library/ AL; **94** Ryan Milne/Sea Mammal Research Unit; **95** Barry Bland/NPL; **96** Green Planet Photography/AL; **97** AA World Travel Library/AL; **98** Bryan and Cherry Alexander/ NPL; **99** Wild Destinations/Polar Regions/ Arctic Alaska/AL; **100** SS; **101** t SS, b piola666/G; **102** Chronicle/AL; **103** SS; **104** t Philippe Clement, b Ocean Science Consulting Limited (OSC); **105** SS; **106** SS; **107** SS; **108** Des Ong/FL; **109** Mike Potts/ NPL; **110** SS; **111** Nick Upton/NPL; **112** AFP/Stringer/G; **113** SS; **114** Hemis/AL; **115** Creative Commons CC0 License; **116** SS; **117** Michael Oats Archives/Stringer; **118** Lofting, H. 1924. Doctor Dolittle's Circus. Frederick A. Stokes; **119** Avpics/AL; **120** t joe daniel price/G, b Universal History Archive/Contributor/G; **121** Jorge Hernando/Dreamstime; **122** t SS, b Stephen Frink/G; **123** Matt Cardy/Stringer/G; **126** Kathy deWitt/AL.

Index